TIME AND THE SPACE-TRAVELLER

TIME AND THE SPACE-TRAVELLER

L. Marder

Senior Lecturer in Applied Mathematics
University of Southampton

University of Pennsylvania Press
Philadelphia

First published in 1971 by George Allen & Unwin, Ltd.

Library of Congress Catalog Card Number: 77–182498
ISBN: 0–8122–7650–7

Printed in the United States of America

Preface

This book was at first conceived as a review of the literature on the clock paradox in relativity theory. The wealth of material which exists on this controversial issue is widely scattered in numerous books and journals, with the result that each time the controversy flares up, the same arguments are put forward with the firm belief that they are original. It seemed desirable, therefore, with the phenomenon of time-dilatation rapidly becoming commonplace (in the laboratory, at least) to gather the material together 'under one roof' and to sort and to examine the arguments in a unified way. The clock paradox has long occupied the attention of the layman as well as the practising scientist, and so it has been my aim to make the book self-contained and to keep the mathematics as simple as possible.

As I read more and more of the literature, it became evident that the central issue – the relativity of time–measurement – needed to be investigated in a wider context if the paradox was not to appear as just a logical parlour game. So often the reader would be urged to 'imagine a space-ship which sets off from the earth at half (or 90 per cent, or 99 per cent) of the speed of light', whereas the imagination of a technologist or physicist might feel this asking too much. I have therefore attempted to deal also with the questions of the limitations of space-travel; the implications of time-dilatation for the long-distance space-traveller; the nature of living clocks; the significance of recent experiments on time-dilatation; the validity of special relativity theory; and so on. Although the treatment is elementary the serious student of relativity (as well as the more general reader with an interest in science) should find the book of interest. For specialist use it will serve, in particular, as a fairly extensive reference source. An annotated bibliography is included.

No apology is made for the inclusion of a rather lengthy historical survey of the foundations of special relativity. This is intended partly for the non-specialist and partly to make precise the concepts and terminology in later chapters.

I am grateful to the many writers who have partaken, over the years, in such an interesting and amusing debate, My special thanks are due to Dr Sebastian von Hoerner of the National Radio Astronomy Observatory, Green Bank, West Virginia; Dr G. J. Whitrow of the Imperial College of Science and Technology, London; the

American Association for the Advancement of Science; and Thomas Nelson Ltd., for permission to include certain material in Chapters 3 and 4. (Full acknowledgements will be found at the relevant places in the text.)

From the start of this study it seemed clear to me which school in the controversy was correct. Nevertheless, the variety of subtle and interesting points which emerged as the work progressed came as a pleasant and welcome surprise.

L. M.

March 1970

Contents

CONTENTS

TIME AND THE SPACE-TRAVELLER

Chapter 1

INTRODUCTION

'All times are not alike.'

CERVANTES, *Don Quixote*

The responsibility of the scientist is vital in many areas of research. Whether he strives to control nuclear fission or fusion, to create life in a test-tube, to transplant organs from one living being to another, or to open the way to space-travel, his search for knowledge can be a hazard to all. Usually the risks are evident and may be calculated. But when scientists disagree over fundamentals, in a theory in constant use, the extent of the hazard is inestimable.

It is therefore, to say the least, remarkable that one of the most controversial issues in modern physics concerns the very basis of Einstein's special theory of relativity, which has long been regarded as one of the most tested theories in the whole of physics. So far, special relativity has survived all its tests, and it is taken for granted daily in thousands of laboratory calculations throughout the world. Yet some of its most startling predictions, concerning the passage of time for long-distance astronauts, are still in dispute. How, then, can one trust the theory so freely?

The controversial question of whether time passes at the same rate for an astronaut as for his earthbound twin brother is a well-known one, and is usually referred to as the *paradox of the twins* or the *clock paradox*. It, and related issues, have led more than one eminent scientist to express doubts about the validity of the special theory of relativity, but none more forcibly than Herbert Dingle, Professor of Natural Philosophy at the Imperial College of Science and

Technology from 1937 to 1946, of History and Philosophy of Science at University College, London from 1946 to 1955, and former President of the Royal Astronomical Society. In an article in *Nature* in October, 1967 [75], Dingle wrote:

> 'Five years ago I gave, as the culmination of several similar efforts, a simple proof that the special relativity theory was untenable. . . . Nevertheless, the theory has continued to be accepted and used as though it were unquestioned. . . .
>
> 'Truth is immortal but human lives are not, and they have claims to protection, even at the cost of admitting an error in physical theory that should never have been made. . . . I hope, therefore, that this matter will no longer be allowed, by neglect, to take its own natural and possibly disastrous course, but will be faced squarely and promptly, with no aim but that of arriving at the truth, whatever it might be.'

Many other researchers have been uneasy over the question of time, in relativity theory, as is evident from the hundreds of papers that have been published on the subject since the formulation of Einstein's theory in 1905, though few would support Dingle in demanding the complete rejection of the theory. Most of the published work is concerned, in one form or another, with the simple issue of whether the astronaut's clock and that of his twin brother would record the same total elapsed time for the astronaut's round-trip journey to a distant star. Conventional relativists believe the times to be unequal, the astronaut's time being the lesser of the two, while most others argue that the twins' clocks would register the same elapsed time. A third and smaller group of theoreticians believe that special relativity predicts both results, and it is this group who find the theory unacceptable through its inconsistency.

Retarded ageing in space-travel is, in any case, insignificant when the speed of the traveller relative to his twin on earth is very low compared with the speed of light, unless the journey is a very long one indeed. But with really high-speed travel the predicted effect can be close to fantasy. Professor G. A. Crocco, then President of the Italian Rocket Society, is reported in *The Times* of September 19, 1956, to have stated, at the 7th International Astronautical Congress in Rome, that 'while centuries passed by travellers would experience only the passage of minutes and would become "almost immortal"'. *The Sunday Times* of September 23 of that year, reporting on the same Congress, refers to a paper 'The Possibility of Reaching the Fixed Stars' by the 'chief German delegate' Professor Sänger, and quotes from it the statement 'Terrestrial years pass like seconds for the crew, who could fly for years without growing a day older'.

In his book, *The Foreseeable Future* [223], Sir George Thomson considers an hypothetical journey at half the speed of light to our nearest star and back. The distance of the nearest star (Proxima Centauri) is about four and a quarter light-years, and the journey would take seventeen earth years, although for the traveller the time would be only fourteen and a half years. This last figure is calculated from the relativistic formula

$$\frac{\text{Traveller's time}}{\text{Earth time}} = \sqrt{(1-k^2)},$$

where k (in this case $\frac{1}{2}$) is the traveller's speed relative to the earth expressed as a fraction of the speed of light. (The period of variable speed during turn-round is regarded as negligible.) Thomson refers to the 'paradox', but says that 'the best opinion is that the contraction [of time] would occur and that the returning astronaut would, in fact, find that time had gone more rapidly on earth than on his space-ship'.

Dingle [55] has violently rejected such suggestions of different ageing for astronaut and earthman. In answer to the statement attributed to Crocco, he commented, 'The otherwise incredible fact emerges that physicists of distinction – men holding high positions in universities and research laboratories – so completely misunderstand the theory of relativity as to think that it does, in fact, entail these fantastic consequences', and warned, 'a situation in which the material destiny of the world is in the hands of man manipulating a tool whose nature they quite misconceive is one of extreme peril'.

How does the question of different ageing rates arise in special relativity theory? The theory is concerned with observations and

13

measurements that are made in *inertial reference frames*,* those reference frames in which, prior to relativity, the Newtonian laws of mechanics were supposed to be valid. Each of these frames comprises a coordinate grid extending throughout space, and it is imagined that a standard clock is conveniently stationed near each point for recording the time of local events.

According to Newton, time is 'absolute'; there is but one time for all observers. Therefore, the battery of real or imagined reference clocks in the universe could, in principle, be adjusted and synchronized to read this time. This precludes the possibility of a clock paradox in Newtonian theory. It goes without saying that Newtonian observers would always agree as to whether two particular events in the universe are simultaneous or not.

On the other hand, fundamental to relativity is the *relativity of simultaneity* – different inertial observers do not always concur on matters of simultaneity. This makes it impossible to devise a universal criterion for synchronizing clocks in relative motion because the synchronization of two clocks is a process of setting them *simultaneously* to read the same. In particular, a uniformly moving clock is not observed to run at the same rate as those at rest in the inertial reference system of the observer concerned; it runs slow by the factor $\sqrt{(1 - k^2)}$, k being the speed of the clock expressed as a fraction of the speed of light. (E. G. Cullwick in his well-known text-book *Electromagnetism and Relativity* has erroneously argued, however, that while receding clocks run slow, approaching clocks run *fast* [45].)

Disputed questions generally begin at about this point. Is the difference in clock rates 'real' or simply a result of somewhat arbitrary comparison procedures? What happens when the 'moving' clock accelerates? Does, in fact, the round-trip space-traveller physically age less than his earthbound twin? And, of course, is there a paradox?

Experiments of ever-increasing variety are adding more and more weight to the opinion that ageing depends on motion. Fundamental particles, *mesons*, produced as a result of interactions occurring in the

---

*Experience shows that the frame in which the 'fixed' stars are at rest is very nearly inertial. Any frame which moves with constant speed in a fixed direction (without rotating) relative to an inertial system is itself inertial, and for most present purposes, the centre of the earth may be regarded as being at rest in one.

upper atmosphere, as well as mesons produced in the laboratory, have 'lifetimes' which are statistically known. The meson's life ends when it 'decays' into other particles, and it is found that the lifetime depends on the speed of the meson, relative to the observer who measures it, in accordance with the relativistic formula. Further confirmation is provided in more recent experiments based on the *Mössbauer effect,* discovered by Rudolph Mössbauer in 1958 (for which he subsequently was awarded a Nobel Prize). This effect permits the production and comparison of gamma rays with extremely well-defined frequencies, which can act as high-precision timekeepers. However, Dingle and some others dispute the interpretation of the various experimental results.

It should be made clear that there are two distinct sides to the problem. The first is concerned with the predictions of a theory. Here is needed no experiment and no elaborate equipment. Indeed, arguments about the time concept and the clock paradox in relativity were in full swing long before experiments capable of detecting relativistic effects were at all feasible. From this point of view, the clock paradox is akin to Zeno's paradox of Achilles and the tortoise.

On the other side is the experimental evidence. If experiments *are* in conflict with theoretical predictions then the theory must fall. But the issue is seldom this simple, as the experiments we *can* perform are rarely the ones we should *like* to perform. In our case space-ships are still hard to come by, and so instead we must make inferences from experiments with mesons and the Mössbauer effect. While most physicists regard the experimental evidence of retarded ageing as extremely convincing, others have been quick to point out what they feel are fundamental differences in principle between the experiments performed and the conditions of space-travel. We shall later review the experimental evidence in some detail (Chapter 5).

The difficulties inherent in comparing the rates and readings of separated clocks have (rightly) been emphasized by Dingle; a few of his opponents have skated over very thin ice in this respect. In his analysis [56] of Thomson's discussion he declares:

'to compare it [the space-traveller's clock] with a terrestrial one you must read them at the same time, otherwise you can get any result you like. The terrestrial observer choosing what he calls the same moment, finds that the traveller's clock is slow. The traveller, however, says that the terrestrial observer has compared them at different moments; he makes the "right" choice and finds the terrestrial clock to be slow. Both are right – which would be impossible if the effect were something that happened to a clock instead of a judgement of simultaneity.'

This agrees with the conventional view as far as the outward part of the journey is concerned. But his conclusion that when the observers meet again at the end of the round trip, and are once again relatively at rest, their judgements of simultaneity agree 'and so their clocks agree' by no means follows. However, Dingle felt that this account was only a 'paraphrase' of Einstein's original paper [83] on the subject (in 1905), although he asserted that Einstein's paper 'contains a most regrettable error, in a statement that a clock moving in a closed curve will be found, on returning to its starting point, to be behind a stationary clock'.

His most persistent opponent in the ageing controversy has been the well-known cosmologist William McCrea, FRS, Research Professor of Theoretical Astronomy at the University of Sussex, who also is a past President of the Royal Astronomical Society. McCrea has dissected and examined Dingle's succession of arguments over the years with patience that has won the admiration of scientists of less stamina. Of the 'paraphrase' he commented 'Dingle's "paraphrase" of Einstein's paper is a travesty', and declared, 'In spite of what Dingle writes, physicists and mathematicians are quite certain about what follows rationally from the postulates of relativity theory'.

Dingle's arguments have almost always been based on the conviction that, in the context of the special relativity theory, there is symmetry in the behaviour of the space-traveller and the earthman except during the three short periods of acceleration when the space-ship sets off, reverses, and is finally brought to rest on earth. Apart from these brief periods, the motion is simply one of uniform relative recession, or approach, of twins. If the earthtwin regards the spacetwin as ageing more slowly than himself on the outward journey, then the converse must be true. (This in itself is not a contradiction, but simply an expression of two different observers' viewpoints.) A similar argument applies on the return journey, and if the twins agree that the periods of acceleration form a small part of the journey, then, on reuniting, they must find each other to have aged an equal amount. Dingle has always regarded the matter as so simple and clear-cut that any intricate mathematical arguments can only be superfluous, and must inevitably cloud the issue. When, following a flurry of correspondence to *Nature* in 1956, the editor of that journal invited 'concluding statements' from the two main participants, Dingle's statement [58] contained a scathing attack on those who allowed themselves 'to be bemused by mathematicians', and ended, 'It is no accident that we [in Great Britain] rank high as builders of mathematical edifices on rotten foundations'.

Earlier in his 'concluding statement', Dingle had placed his opponents into the two classes '(a) those who held that accelerations are irrelevant and that Dingle misunderstands special relativity; (b) others'. It is clear that McCrea falls into class (b), because McCrea insists that though the time actually spent in acceleration may be negligibly short, the *effect* of the periods of acceleration is by no means negligible.

It is hardly surprising that such a controversy on space-travel evoked the interest and participation of the public at large. By April, 1956, the debate had spread beyond the confines of the learned journal of the professional scientist. On the 29th of that month, the following article by John Davy, Science Correspondent, appeared in *The Observer* (reproduced in full by kind permission of that newspaper):

CAN SPACE TRAVEL POSTPONE DEATH?

Could high-speed space-travel keep you young? This seems to be the crux of a controversy which occupies some space in yesterday's issue of *Nature* (writes John Davy).

Many scientists have concluded from Einstein's theory of relativity that time 'contracts' when a body moves, so that, as Sir George Thomson said in his recent book 'The Foreseeable Future', an astronaut returning from a journey lasting seventeen terrestrial years would have aged by only fourteen and a half years –

i.e. time would have gone more rapidly on earth than in his space-ship.

Professor Herbert Dingle has now written a vigorous article saying that this would not happen, and that Einstein's theory is being misunderstood. The essential point of Professor Dingle's argument appears to be this: Einstein pointed out that we have no absolute way of discovering if two events happening in different places are simultaneous. Thus, if someone on the earth and an astronaut out in space try to compare their watches at what they think is the 'same' moment, each will think that the other's watch is slow – and both will be, in a sense, right.

But when the astronaut returns to earth, and watches are checked, again at the 'same' moment, their judgment of what is the same moment will coincide, and the watches will agree. 'All this,' says Professor Dingle, 'applies also to heartbeats. The observers will have "lived" the same time and made the same progress towards the tomb'.

Professor Dingle's conclusions are disputed by Professor McCrea, of California University [where he was Visiting Professor], who says his 'exposition of relativity theory is wrong'.

One week later, in *The Observer's* correspondence columns, Professor William Wilson, FRS, attempted to refute the relativistic prediction of differential ageing. He argued that the astronaut's time reckoning could be used for events on the earth, and that the terrestrial time reckoning could be used for events in the space-ship, and concluded that in either case the returning astronaut would have aged by precisely the same amount as those whom he left behind on earth. Analogously:

'A fahrenheit thermometer gives the temperature of the water in a bath as 40 degrees; on replacing it by a centigrade thermometer we find the temperature to be 8 degrees! But surely no sane person imagines that this interchange of thermometers has cooled the water.'

(The arithmetical error in conversion was subsequently pointed out in correspondence by 'Arenceste, of Form IV' with the wry comment:

'We have been getting this sort of answer for years, but nobody ever believed us'.) But it is hard to imagine that the controversy would have reached its eventual proportions, had the answer been as simple as propounded by Professor Wilson.

At about this time, public interest reached its peak. Letters to correspondence columns of newspapers expressed viewpoints that were varied, sometimes irrelevant, but hardly ever dull. Typical was the following delightful communication appearing in *The Observer* of May 6, 1956, from a Major F. A. Yorke of Bournemouth:*

'Sir, – Pundits have always hedged on this particular, if rather foolish, conundrum. As far as I see it, the whole problem depends on the fact that the astronaut is, on departure, subjected to a vast but artificial gravitational field due to the acceleration of his space-ship against natural gravity. Such an artificial field along the vector direction of drive would slow-down or set-back the rate of change of all things movable within the ship, from the processes of heart, lungs, cells and, more especially, brain-cells, to the 'innards' of watches and other instruments. For compactness let us call these rates of change metabolism.

As soon as the power-drive is cut, the ship will continue with the same velocity, but now uniform. The British term for this is 'coasting', though the Americans prefer 'free-fall', a definition which suggests an impending crash and which I think to be – apart from missiles – misleading.

During uniform motion or coasting there is nothing to interfere with the newly adopted metabolism, and the astronaut, if still surviving, will be perfectly happy with his lot. When at length he comes in to land on earth, he will have to apply the same force to decelerate his ship as he utilized at departure: the debt must be repaid – I would say with interest, for now metabolism would race forward to the state normal to earth.

The probable result would be that, after comparing watches to mutual satisfaction, the astronaut would gradually pass out, a victim to katabolism.'

*Reprinted by kind permission of *The Observer*. Unfortunately, neither that newspaper nor myself was able to trace Major Yorke to seek his own permission.

Though no doubt written with tongue-in-cheek, Major Yorke's letter focuses attention on two questions which frequently arise. The first is, to what extent is a human being a clock? Most physicists (though there *are* one or two notable exceptions) believe that the human body, by means of its heartbeats, cell growth and division, and other life processes, keeps time in step with a 'normal' clock in the same state of uniform, or *moderately* accelerated motion, whether that clock be controlled by balance wheel, quartz vibration or gyromagnetic resonance (as in the *caesium clock*). (Very high acceleration, as might be due to unduly powerful rocket motors, will of course damage both human and other clock mechanisms.)

The second question, closely linked with the first, concerns the traveller's bodily *awareness* of the new time-rate following a change of motion. If a transition is made from one state of inertial motion to another, is the resulting 'change' in metabolism something to which the body needs to adjust, like a change in ambient temperature? The true relativist would emphatically say no. But let us quote one dissenter to the common view. L. O. Pilgeram, of the Arteriosclerosis Research Laboratory, Minneapolis, has said [189] in criticizing work of S. von Hoerner (see p. 98) that Von Hoerner makes 'unwarranted assumptions when he attempts to apply Einstein's relativity theory to biological time', and, 'There is no known causal means by which greatly increased velocity could alter, without destroying the very biochemical basis of the life process, those metabolic changes which are responsible for the ageing process'.

Differential time-rates in relativity were first mentioned in Einstein's famous 1905 paper, referred to earlier, in which he formulated his special theory. The subject was discussed in detail in an article [151] by Paul Langevin in 1911, in which 'Langevin's traveller' (who corresponds to our astronaut) was introduced. Langevin was in no doubt that differential ageing was in accord with special relativity, but his work caused much subsequent controversy, especially in the 1920s. (Some of these early conflicts have been amusingly reviewed by Henri Arzeliès [2].) But at that time, and until comparatively recently, the issues were largely academic, concerning logical deductions in relativity theory. With the advent of ambitious space journeys these same issues may become of direct concern to the astronaut.

A detailed relativistic analysis of the limits of space-travel, taking into account limitations imposed by technological and biological factors, was made by von Hoerner in the early 1960s while at the Astronomisches Rechen-Institut, Heidelberg, Germany [230; 231]. He first estimated from a number of premises the probability that any particular star (chosen at random) would be the centre of a solar system which supports intelligent life. His estimate suggests that the ten nearest such stars are at an average distance of rather less than twenty light-years from us. Next, von Hoerner surveyed the present and future possibilities of travelling this far. It is unlikely that man could stand accelerations much in excess of $1g$ for periods of many years (although $2g$ or $3g$ might prove tolerable), and so the astronaut cannot travel the whole way at speeds comparable to that of light, in the manner of Sir George Thomson's traveller. Nevertheless, if a constant acceleration of $1g$ is maintained for the first half of the outward journey and an equal deceleration is maintained for the second half, the astronaut will just come to rest at the distant system, having reached when half-way out a maximum speed of about 0·995 times the speed of light. (The calculation is performed relativistically.) If the return journey is made in the same way, the total travel time is about forty-two terrestrial years.

In calculating the astronaut's time of travel the *clock hypothesis,* first made by Einstein [*ibid*], was employed; this states that an accelerated clock agrees in rate with an unaccelerated one which is momentarily moving alongside at the same speed.* The astronaut's clocks are therefore found always to be running slow in the inertial system of earth by the factor $\sqrt{(1-k^2)}$, which is now variable. The full calculation shows that the whole journey takes only twelve years for the astronaut, representing the considerable saving of thirty years. (The practicability of the journey is considered in Chapter 3.)

An attempt is made in this book to examine a wide range of issues in the clock paradox controversy, including: the basis of special

---

*The clock has to be of an approved type. An atomic clock is suitable, but one which depends on gravity, like a pendulum clock, is not; acceleration affects the apparent gravitational field.

relativity theory; the problem of biological time; the prospects for long-distance space-travellers; the many logical arguments on both (or should one say *all*?) sides in the central dispute; and the significance of some recent experiments. And because many authors believe that the paradox can only be resolved by appeal to the general theory of relativity, an account of it is given in the context of that theory.

The literature on this subject is prodigious. Its study leads me firmly to believe, in common with most physicists, that differential ageing is a feature of the world we live in, though equally firmly I believe that many will continue to dispute this. The account which follows is aimed, as far as possible, at both the specialist and non-specialist reader. The latter may, perhaps, be most interested in the more 'spectacular' consequences of the ageing phenomenon; it is hoped, too, that he will find reassurance as to the soundness of a theory on which so many vital predictions are already based.

Chapter Two

BACKGROUND TO RELATIVITY

1. Don't bring back the ether!

It is sometimes said that Einstein's rejection of the concepts of absolute space, absolute time and the 'all-pervading' light-transmitting medium called the *ether*, or *aether*, was not dependent upon the convincing failure of optical experiments in the last century aimed at detecting that medium. By this assertion, experiments like the famous one of Michelson and Morley performed in 1887 (see p. 27) to detect the earth's motion through the ether, were not crucial to the eventual emergence of the *principle of relativity*. The principle of relativity, which states that physical laws are the same in every inertial reference system, finds no place for a preferred system such as that which would be determined by the ether. All uniform motion is therefore relative, and absolute space and an absolute state of rest become meaningless notions.

In his autobiographical notes, Einstein [257] has said in reference to the need for the radical new principle:

'By and by I despaired of the possibility of discovering the true laws [of mechanics and thermodynamics] by means of constructive efforts based on known facts. The longer and more despairingly I tried, the more I came to the conviction that only the discovery of a universal formal principle could lead us to assured results ... How, then, could such a universal principle be found? After ten years of reflection such a principle resulted from a paradox upon which I had already hit at the age of sixteen: If I pursue a beam of light with the velocity c (velocity of light in a vacuum), I should observe such a beam as a spatially oscillatory electromagnetic field at rest. However, there seems to be no such thing . . . From the very beginning it appeared to me intuitively clear that, judged from the standpoint of such an observer, everything would have to happen according to the same laws as for an observer who, relative to the earth, was at rest.'

Although Einstein did make reference to the ether in his autobiographical notes, and also in his 1905 paper, where he stated 'the unsuccessful attempts to discover any motion of the earth relatively to the "light medium" suggest that the phenomena of electrodynamics as well as of mechanics possess no properties corresponding to the idea of absolute rest . . .', it seems evident that the relativity principle evolved, in his own mind, from rather more general considerations.

In *Michelson and the Speed of Light,* Jaffe [270] refers to a passage in a letter he had received from Einstein, tending to confirm this:

'It is no doubt that Michelson's experiment was of considerable influence upon my work insofar as it strengthened my conviction concerning the validity of the principle of the special theory of relativity. On the other side I was pretty much convinced of the principle before I did know this experiment and its results.'

Einstein was not, however, the first to put forward the principle of the relativity of uniform motion. In 1904, Poincaré [289], having previously provided the groundwork, described the principle and even gave it the name 'principle of relativity'. Sir Edmund Whittaker, in his excellent treatise *History of the Theories of Aether and Electricity* [304] speaks provocatively of 'the relativity theory of Poincaré and Lorentz', and describes how, in the autumn of 1905, 'Einstein published a paper which set forth the relativity theory of Poincaré and Lorentz with some amplifications, and which attracted much attention'. Other writers (e.g. Holton [268]) have subsequently criticized Whittaker for inadequately appreciating Einstein's role in the development of the theory. We shall return to this point later.

If we follow Einstein's expressed way of thinking, a detailed study of the ether concept is by no means essential for the understanding of relativity theory. Nevertheless, we shall find it worth while, in this section, to review briefly the turbulent history of that concept because it plays a significant part in a number of relevant discussions. (For fuller details, see Whittaker [303] or Hesse [266].)

To the Greek scientists, the ether meant simply the clear sky, beyond the air and clouds which surround the earth. The idea of a substantial ether, with mechanical properties, was put forward by René Descartes (1596-1650). Descartes found the notion of a void unacceptable, and held that spatial extension and substance were inseparable because every perception of space involves bodies and, conversely, bodies are essentially shapes with motion. Action at a distance, apparently experienced by magnets, and by the seas in the tidal motion which was observed to depend on the position of the moon, was impossible. The space between bodies, and between the constituent particles of bodies and the atmosphere must, thought Descartes, be completely filled with a fluid medium comprising particles in contact.

The essential factor in respect of action over distances was that the action would be transmitted from point to point through space by means of ether particles in contact with one another. The motion of any body resulted in the motion of ether particles, but only complete rings of particles could move because otherwise empty spaces would be created, which was impossible. In all, Descartes envisaged the existence of three kinds of matter of which the ether particles were one, the other kinds being luminous particles as in the sun and stars, and opaque reflective matter like that of the earth and other planets. Light was transmitted by pressure between the particles of ether, while *vortices* formed of the luminous type of matter played a major part in magnetism, electricity and gravitation.

All this was very grand and ambitious, but it was also unscientific, because his ideas were largely arbitrary and sufficiently vague to fit almost any observed phenomena. His work was widely attacked, though not quickly discarded. Two main rival theories of light came to be put forward; the corpuscular theory, by which light consists of corpuscles emitted by a luminous body, and the wave theory, which assumed light to be propagated as undulations in an elastic medium. At first, only longitudinal waves, akin to those of sound, were considered.

Newton's famous observations of the refraction of sunlight by prisms heightened an interest in the ultimate nature of light. Largely because of the inability of the longitudinal wave theory to account for phenomena subsequently identified with the polarization of light, Newton denied that ether vibrations were responsible for light transmission, and favoured the corpuscular hypothesis. Over a century later, Young's investigations of interference and refraction showed the superiority of the wave theory, and the optical studies of Young and Fresnel (especially on polarization, during the years 1817 to 1827) led to the eventual firm acceptance of the wave theory of light, in which the oscillations are transverse and not longitudinal motions of the ether.

How was the ether itself pictured? In the nineteenth century, all manner of possibilities were considered. A gaseous ether could support longitudinal, but not transverse waves, and so the ether was described as an elastic solid. Some believed it to be fairly rigid, others thought only a highly tenuous material could allow the free motion of ordinary bodies. Some thought that the properties of the ether were different in different circumstances, or that there were a number of ethers, coexisting for different purposes. An intensive study of waves in elastic solids was proceeding at that time, and the problem of the absence of longitudinal optical waves was studied by MacCullagh, who devised an ether theory based on a medium which resisted, elastically, only the *rotation* of its elements. This 'rotationally-elastic' theory subsequently proved one of the most successful.

Faraday's experiments with electricity and magnetism in the first half of the nineteenth century, and his distaste for action at a distance, impelled him to introduce his ideas of physically real electric, magnetic and gravitational lines of force filling the space between particles. These might even constitute the ether. Of the more mathematical workers in electromagnetism, Maxwell, partly influenced by Faraday, introduced a mechanical model of the electromagnetic field in which electric effects were linear and magnetic effects rotary. But the precise nature of Maxwell's ether was unclear, although his electromagnetic equations were to prove so brilliantly successful.

In the early 1860s, Maxwell was able to deduce that electromagnetic disturbances travelled with the velocity of light, which had been calculated from observations of Jupiter's moons as early as 1676 by Roemer, a Dane. By the 1860s the more accurate terrestrial measurements of Fizeau and Foucault were also available. Thus, Maxwell identified electromagnetic radiation with light, and the electromagnetic and luminous ethers became one and the same.

It was only after detailed investigations of the effect that a moving body might have on the ether within and around it that the eventual demise of the ether theory came. By then, more and more ingenious models involving vortices, *vortex-rings* and *vortex-sponges* had been considered. But little of the mechanism of some of the mechanical models described would actually be observable as an electromagnetic field. Otherwise, as Born later said, 'the ether would be a monstrous mechanism of invisible toothed wheels, gyroscopes and gears intergripping in the most complicated fashion'. Thus, the abandonment of the ether theory, with the introduction of relativity, came as a relief to many.

In October, 1967, when the cases against and for special relativity were being argued by Dingle and McCrea in *Nature,* the editor of that journal summarized the discussion [183]. Perhaps it was with genuine concern that he entitled his editorial 'Don't Bring Back the Ether'.

2. Michelson-Morley and all that*

One day in 1879, the local newspaper of Virginia City in Nevada proclaimed that 'Ensign Albert A. Michelson, a son of Samuel Michelson, the dry goods merchant of this city, has aroused the attention of the whole country by his remarkable discoveries in measuring the velocity of light'. Michelson, whose life was to be devoted to the ultimate precision in measurement, was the first American to succeed in determining the speed of light emitted by a

*Biographical details of Michelson are to be found in Jaffe's book, *Michelson and the Speed of Light* [270].

terrestrial source. By 1882 his experimental technique was refined to the extent that his new value, 299 853 kilometres per second, for the speed of light in a vacuum was accepted for forty-five years, until he again improved on his accuracy.

Much of Michelson's life was spent in the design and construction of spectrographic apparatus. He made his own diffraction gratings, and these were required to be so precise that he spent many years constructing the machines with which to make them. Some of his apparatus was used for such purposes as measuring the apparent diameters of stars (which telescopes alone could only detect as points) as well as in determining the light speed. But he is probably known best of all for his experiments aimed at detecting the flow of the ether past the earth.

In 1881, as a professor on leave from the Case School of Applied Science in Ohio, he was at the University of Berlin, where he attempted his first ether measurement [280] using a newly-devised *interferometer*. The vibration of his apparatus due to traffic caused him to make a second attempt at an observatory in Potsdam, and it was here that he first came to the remarkable conclusion that there was no ether flow past the earth; i.e. that *the earth was at rest in the ether*.

Another American, Edward Morley, who was a chemist, became interested in Michelson's work and joined him in a repeat experiment with improved apparatus in 1887 [281]. Michelson was aware of deficiencies in his 1881 experiment, and the two were to make sure of things this time. The principle of the Michelson–Morley experiment is shown schematically in Figure 1. A light ray emitted by a mono-chromatic source S arrives at a thinly-silvered mirror M_1, and part of the light passes through M_1 to a second mirror M_2, where it is reflected back to M_1 and thence to the telescope T. This part of the original ray is called ray 1. The remainder of the light (ray 2) from source S is reflected through a right-angle by the mirror M_1, and reaches another mirror M_3, which returns it back to M_1. Some of this light then passes through the mirror M_1 and travels into the telescope T. If the distances M_1M_2 and M_1M_3 are the same, each being equal to l, say, then the total distances travelled by the light in rays 1 and 2 are equal.

Suppose first that the apparatus is at rest in the ether. Then the speed of light relative to the apparatus is the same for the two rays, which consequently reach the telescope together. But the detailed nature of the reflection at M_1 is not quite the same for the two rays, and as a result the oscillations of rays 1 and 2 are not in phase at the telescope T, causing the rays to *interfere*. By making fine adjustments to the mirrors the observer is able to see this interference through the telescope as alternately placed light and dark bands.

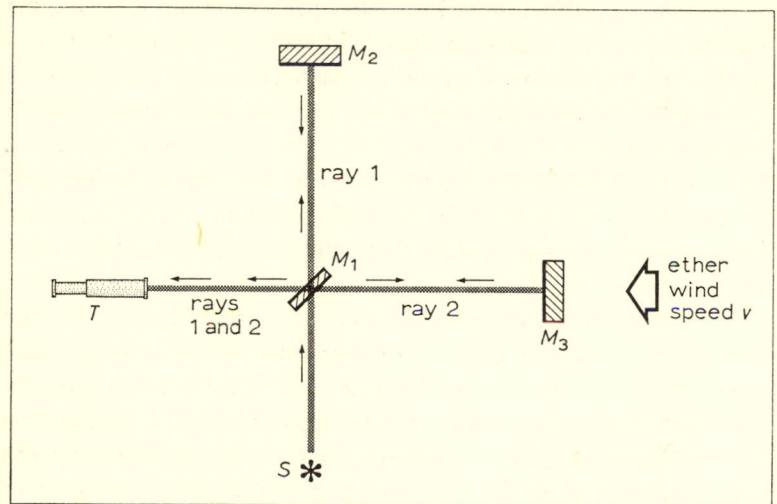

Fig. 1. *The Michelson-Morley experiment*

Next, let us imagine that the ether is flowing past the apparatus, because of the earth's motion, and that the ether wind has speed v in the direction M_3M_1. Then the light speed for ray 2, relative to the apparatus, is $c+v$ when travelling in the direction M_3M_1 and $c-v$ when travelling in the opposite direction. The speed of ray 1, relative to the apparatus, will be affected differently. The rays no longer take the same time to travel from source to telescope, and therefore the interference pattern is not the same as when there is no ether wind.

Essentially, the experiment consists in determining the ether wind velocity by observing the change in interference pattern as the whole apparatus is rotated into different orientations. For example, a rotation through 90° in the case considered in Figure 1 interchanges the roles of rays 1 and 2, and the time *difference* of travel for the two rays changes by approximately $2lv^2/c^3$ seconds. The observed change in pattern enables the unknown value of v to be found.

Thousands of repeated measurements over a period of time and with all orientations of the interferometer were made by Michelson and Morley. On the assumption that the speed of the earth through the ether was of the same order of magnitude as its orbital speed around the sun (about thirty kilometres per second) for at least some part of the year, a clearly observable effect was expected. None was found. Subsequent experiments by many physicists have similarly failed to detect an ether wind, except for an unexplained positive result by Dayton C. Miller in 1925 [282]. Recent experiments, using *maser* (Microwave Amplification by Stimulated Emission of Radiation) beams, or the Mössbauer effect (see p. 154), are especially important as these are capable of detecting an ether velocity as small as a few metres per second.

There were numerous attempts to explain the Michelson–Morley experiment on classical grounds. Of particular interest is the idea that a moving body drags the neighbouring ether. Fresnel, much earlier in the nineteenth century, had suggested that the ether in a refracting medium such as glass is partially dragged, to explain why the refraction of starlight by a prism was found to be independent of the apparent direction of the star, and others considered complete dragging of the surrounding ether by the earth. The two suggestions were, of course, contradictory.

A quite different explanation of Michelson and Morley's observations was that, as suggested by Ritz, the light velocity is c *relative to the light-source* and not relative to the ether. Thus, experiments of the Michelson–Morley type using terrestrial sources would always have negative results. On the other hand, light from an approaching star would travel towards the earth more rapidly than light from a receding star, and this should show up as apparent irregularities in the orbits of certain double stars (which comprise pairs of stars rotating about each other under their mutual gravitational attraction). De Sitter's observations [254] of double stars in 1913 failed to reveal these predicted irregularities.

Lorentz was able to account for the result of Fresnel in his 'theory of electrons', which attempted to deal with electromagnetism at the microscopic level. In Lorentz's theory, the electrically charged constituent particles of bodies acted on one another via the ether which, however, was not dragged at all but was quite immovable. On the assumption that Maxwell's equations were valid in the 'absolute' reference frame of the ether, Lorentz developed his theory a considerable way. However, it did not fit all the facts, and it could not account for the Michelson–Morley result without the introduction of radically new ideas.

Revolutionary suggestions about the kinematical properties of the universe came forward towards the end of the nineteenth century. In June, 1892, Sir Oliver Lodge [274] reported on a hypothesis of Fitzgerald, that bodies moving through the ether with speed v were contracted in the direction of their motion by the factor $\sqrt{(1 - v^2/c^2)}$. This helped to overcome many previous difficulties. In particular, the length of an arm of the Michelson–Morley interferometer, when parallel to the ether wind, would be shortened by just the amount necessary to equalize the time of travel of the two light rays. The irritating experiment was thus explained by a very slight contraction of the apparatus as the ether blew past it.

Shortly afterwards, Lorentz proposed the same contraction of bodies. In 1904, he obtained [277] the equations which relate coordinates and time measurements in the ether and any other inertial frame, if Maxwell's equations are assumed to apply equally in the two frames. These transforming equations, different from those of conventional theory, were subsequently called the *Lorentz transformation* by Poincaré. They were later to be found identical in form to the transformation equations between any two inertial frames, as derived in Einstein's 1905 paper. (See §6.) In Lorentz's equations the times t in the ether system and t' in the moving system are not equal but the time t' was regarded by Lorentz as simply a mathematical convenience; 'real' time was still the absolute time of the ether system. The lack of tangible properties of Lorentz's ether rendered it of little physical significance. It was simply a 'preferred' reference system in the universe, defining a particular state of motion which, if we wish, we may call absolute rest.

Whittaker [304] has emphasized that by the end of the last century Poincaré had expressed doubts on the existence of the ether on more than one occasion, and had stated that optical as well as mechanical phenomena probably depended only on relative motions of the bodies concerned in them. He refers also to a Congress of Arts and Science at St Louis, USA, on September 24, 1904, where Poincaré introduced the 'principle of relativity' according to which 'the laws of physical phenomena must be the same for a "fixed" observer as for an observer who has a uniform motion of translation relative to him: so that we have not, and cannot possibly have, any means of discerning whether we are, or are not, carried along in such a motion'. Poincaré further suggested that a consequence would be a new kind of dynamics characterized by the rule that no velocity can exceed the velocity of light.

Was Einstein's role in the construction of the special theory of relativity as Whittaker tends to suggest, merely a kind of tidying–up operation? A different view from Whittaker's is taken by the physicist and historian of science Gerald Holton [268]. According to Holton:

> 'The paper by Poincaré of 1904 which Whittaker cites turns out not to enunciate the new relativity principle, but is rather a very acute and penetrating though qualitative summary of the difficulties which contemporary physics was then making for six classical laws or principles, including what is in effect the Galilean–Newtonian principle of relativity.'

(The Galilean-Newtonian principle of relativity asserts the equivalence of inertial reference systems for purely *mechanical*, but not necessarily other, phenomena.) But Holton clearly appreciates that Poincaré saw the need to construct a whole new mechanics that, in Poincaré's words, 'we only succeed in catching a glimpse of, where inertia increasing with the velocity, the velocity of light would become an impassable limit'.

Poincaré's 'glimpse' was undoubtedly far-sighted, but his ideas were not yet at the stage at which he could exploit them, whereas Einstein, in the following year, fully exploited his own, independently reached, ideas on the relativity of motion.

3. *The invariable speed of light*

Einstein based his special theory of relativity on two postulates. As he expressed it:

> 'the same laws of electrodynamics and optics will be valid for all frames of reference for which the equations of mechanics hold good. We will raise this conjecture (the purport of which will hereafter be called the 'Principle of Relativity') to the status of a postulate, and also introduce another postulate, which is only apparently irreconcilable with the former, namely, that light is always propagated in empty space with a definite velocity c which is independent of the state of motion of the emitting body.'

The first of Einstein's postulates, the principle of relativity, we have already discussed at length; we shall now be concerned with the second. What we are asked to accept, in this second postulate, may be illustrated by a simple example. Imagine that with very sensitive apparatus we, on earth, try to measure the speed of light which arrives from a distant star. In our experiment we follow the course of a particular identifiable pulse of the light, such as might be emitted during a slight irregularity in the behaviour of the star. The pulse can, in principle, be timed accurately over the measured distance between two parts of our apparatus and hence the light speed is determined by a trivial calculation.

Next, suppose that a space-ship is travelling at high speed towards the star, and that in the space-ship is carried apparatus like our own, constructed in the same factory as ours and identical in every way. Further, let the space-ship carry an exact replica of our timing device, the two timing devices also being supplied by one manufacturer. In the space-ship, the speed of light emitted by the star is determined in an experiment just like that carried out on earth. Then, according to Einstein's postulate, the measured speed of light will be found to be precisely the same in the two experiments. Furthermore, a repeat experiment with light from a different star, in any state of motion whatsoever, will result in the same numerical value for the measured light speed both on earth and in the space-ship.

What direct evidence is there for this proposition? At the time the theory of relativity was being formulated there was remarkably little. The Michelson–Morley experiment could certainly be explained on the assumption that light is propagated with the same speed, c, in every inertial reference frame and hence in the frame of the interferometer. But it could also be explained in other ways, such as by the contraction hypothesis of Fitzgerald and Lorentz, as we have seen. And because the Michelson–Morley light source was fixed to the platform which carried the mirrors and telescope, the negative result was equally compatible with the simpler hypothesis that the speed of light is constant relative to its source.

Another difficulty in drawing conclusions about the light speed from this experiment is that the two light rays, rays 1 and 2, whose travel times are in effect compared, both make two-way journeys. Each undergoes reflection at a mirror, M_2 or M_3, and so it is only the average speed over a two-way journey, to and from the mirror, that enters the experiment. This is a crucial fact, which has been pointed out by many writers. There are always inherent difficulties in any experiment in which the *one-way* speed of light is to be measured, because the timing has to be made at two points in the light path and this either involves the use of separated clocks (which must first be synchronized) or it involves some equivalent procedure. The theoretical problem of the synchronization of clocks will be discussed in the next section.

Inasmuch as the work of Michelson and Morley is connected with the speed of light, the most that could be deduced from their experiment is that, in the words of H. P. Robertson [292], 'The total time required for light to traverse, in free space, a distance l and to return is independent of its direction'. The point here is that by rotating the interferometer one simply changes the *direction* of the light rays, and not the *motion* of the apparatus. The change in motion of the earth (as it progresses in its orbit) was used, more directly than it had been by Michelson and Morley, in an experiment performed by R. J. Kennedy and E. M. Thorndike [140] in 1932. These workers employed an interferometer with unequal arms, and made observations to see whether there was any change in the interference pattern in the telescope as the motion of the earth changed over a period of time.

There was none. Their result could not be explained on the basis of contraction of the apparatus due to motion through the ether, and was concerned with the relativity of time rather than length. The interpretation of Robertson, in this case, was that, 'The total time required for light to traverse a closed path in [an inertial reference frame] S is independent of the velocity v of S relative to [the rest, or ether, frame] Σ '. Thus, we are again concerned with other than one-way light paths, but the comparison now involves the speed of light in a succession of different inertial frames. However, the Kennedy–Thorndike experiment came long after the birth of the theory of relativity.

It seems, therefore, that Einstein's second postulate was made on the supposition, based on a variety of considerations, that the laws of electrodynamics were valid in every inertial frame. It is, in fact, known that the general validity of Maxwell's equations for free space implies the principle of the constancy of the velocity of light.*

Some recent experiments provide the most direct evidence that the light speed does not depend on the motion of the source. De Sitter's analysis of double star orbits had the shortcoming that there might be overlooked effects due to the presence of atmospheres surrounding the systems. The new experiments involve high-frequency radiation (γ-rays) emitted from fast-moving π *mesons*. In 1935, Yukawa had predicted the existence of electrically-charged particles with mass *intermediate* between those of the electron and the proton (hence the name 'meson') to explain the binding forces of atomic nuclei. C. F. Powell and his collaborators at Bristol [273] found evidence in 1947 of these π mesons, in photographic emulsions exposed at mountain altitudes, and not long afterwards uncharged (i.e. electrically neutral) π mesons were detected in cyclotron experiments. Neutral π mesons have extremely short lifetimes, of the order of 10^{-16} seconds, after which time they 'decay' into γ radiation.

In the light speed experiments, four workers [244] at Geneva used γ rays from the decay of π mesons produced in the CERN proton synchrotron. They measured the speed of the π mesons to be about

---

*See, for example, the proof given in Rosser's book, *An Introduction to the Theory of Relativity* [199].

99·75 per cent of the speed of light. Nevertheless, the γ radiation from these very fast-moving sources was found to be equal to c to within an accuracy of one-hundredth of 1 per cent. The mesons' enormous speed had therefore made not the slightest difference to that of the resulting γ radiation. It should be noted that the difficulties involved in one-way speed measurements do not arise here, as in principle the motion of the γ rays can be compared with that of 'ordinary' light over the same laboratory path using the same timing technique.

4. Measuring length and time

It will be useful to review here how measurements of distance and of time are made and what standards are used. For everyday purposes, distances on earth can be measured by placing end to end a number of measuring rods of a suitably rigid material cut to the length of a pre-chosen standard, such as the standard metre. This is essentially the method of the surveyor for short distances. Subdivisions on the measuring rods enable us to measure, also, fractions of the standard unit.

For complete precision in this method we should require 'absolutely rigid' measuring rods. Unfortunately there are no such things; for one reason, there is no 'absolute' criterion of rigidity. It is first necessary to *define* what is meant by rigidity or by constancy of length, although intuitive notions of these concepts limit our choice of definition to a large extent. Even these intuitive requirements are not satisfied by any known material.

Take any pair of familiar steel rules, such as can be purchased in a hardware store. Apply stresses to one of them, or heat it over a flame, and it is no longer the same length as the other. Therefore, ordinary steel cannot be an absolutely rigid material according to any reasonable criterion.

[*Problem:* A steelmaking firm produces 'brand X' rods of 'new improved steel', which defy all efforts to change the length of one rod relative to another, and so satisfy one obvious condition for rigidity. The firm's biggest rival then produces 'brand Y' rods of a slightly different 'new steel' and these, too, defy all efforts to change the length of one rod relative to another. In some weather conditions a 'brand X' rod is found to be longer than a 'brand Y' rod, although in other conditions the reverse is true. How do we decide which brand of rod, if either, to call rigid?]

In such matters an arbitrary standard has to be chosen. From 1889 until 1960, the standard metre was defined to be the distance between two fine marks on a particular platinum-iridium bar, housed at the International Bureau of Weights and Measures near Paris, when the bar is at the temperature of melting ice. Copies of this bar, *secondary standards*, were then used in the calibration of other forms of measuring instrument.

During the time this standard was in use, experiments showed that the wavelengths of certain atomic spectral lines were constant to a high degree of accuracy. A definite number of wavelengths of the radiation, therefore, occupied one metre. These spectral lines could be reproduced in laboratory equipment constructed anywhere in the world (or elsewhere) provided that the necessary materials were available, and could be used for rather accurate measurement of distances; it would not be necessary to calibrate each piece of equipment against a standard metre bar. This feature is highly convenient and so in October, 1960, it was decided at the Eleventh General Conference on Weights and Measures, that the metre would henceforth be defined optically, with reference to a particular spectral line. The metre is now, by definition, 1 650 763·73 times the wavelength *in vacuo* of the orange-red spectral line in krypton 86.

The measurement of time is, likewise, a comparison procedure. What is actually measured is a *time-interval* between two events, and this measurement has to be made by comparison with some repetitive process chosen as a standard. To be satisfactory, the standard must satisfy some preconceived ideas of regularity. For example, in everyday use, clocks governed by pendulums of the same length agree with each other fairly well, and by these clocks the days and years appear approximately of constant length, human pulse rates are generally steady, and objects dropped under similar conditions take about the same time to fall to the ground. Pendulum oscillations could therefore be used in the choice of a simple form of standard.

As a more satisfactory astronomical standard, the *sidereal day* has been much used [301]. This is the time taken for the earth to make a complete revolution, the measurement being made by reference to the direction of a hypothetical fixed star, known as the First Point of Aries. Sidereal time is not convenient for everyday practical purposes, because the sidereal day is about four minutes shorter than that determined by the apparent position of the sun, and so daylight hours do not always occupy the same part of the sidereal day. (The four-minute discrepancy is due to the earth's motion in its orbit.) The time measured by the apparent position of the sun is called *solar time*, and the sun crosses the Greenwich meridian at intervals of one *solar day*. Mainly because of the tilt in the earth's axis, the solar day is of variable length when compared with the sidereal day, and for precision purposes it is necessary to average out the variation. This leads to the definition of the *mean solar day*, which is the average length of the solar day in terms of sidereal days and is the basis of civil and legal time.

In 1939, it was found by Harold Spencer Jones (Astronomer Royal from 1933 to 1955) that mean solar time and sidereal time are both different from the time which applies in Newtonian mechanics [300]. Spencer Jones made a very detailed analysis of the observed motions of the sun, moon and planets, and found that these could all be explained if, in Newton's equations of motion, the time variable t progressed non-uniformly with respect to the other time standards. The amount of non-uniformity was, of course, extremely small and hard to detect. But Newtonian time is of such fundamental importance in astronomy in predicting the positions of the planets and in other calculations involving dynamical laws, that it was eventually decided, at the General Assembly of the International Astronomical Union in Rome, in 1952, to adopt a new time called *ephemeris time* based on the Newtonian variable. Ephemeris time would be measured from a definite instant in the year 1900, and the unit of time would be the sidereal year 1900. Four years later, the International Committee of Weights and Measures adopted the ephemeris second as a fundamental time unit, to be defined as 'the fraction 1/31 556 925·9747 of the tropical year 1900'.

Ephemeris time has one major shortcoming. The epoch (the word means time measured from a definite date) of an event is determined precisely only in retrospect. Some years of astronomical observation of the solar system is necessary to fix, with precision, the progress of the Newtonian time variable, and hence the ephemeris time.

In the measurement of short time-intervals, the frequency of the radiation in atomic or molecular processes can be used. In 1967, the General Conference on Weights and Measures agreed to replace the definition of the second by

'9 192 631 770 periods of the radiation corresponding to the transition between the two hyperfine levels of the ground state of the atom of Caesium 133.'

Atomic time is particularly suitable for use in the theory of relativity, when reference is to be made to identical clocks at rest in different inertial frames, because, unlike ephemeris time, it is not based directly on observations made on earth. The President of the Royal Astronomical Society, D. H. Sadler, in his Presidential Address [294] on February 9, 1968, 'Astronomical Measures of Time', commented on the new time unit and the way it had originated:

'There can be no question that the new unit is greatly superior to the second of ephemeris time for all purposes other than fundamental astronomy. . . . The new unit is now reproducible from off-the-shelf commercial apparatus with a precision of about 1 part in 10^{11}; and, as has been mentioned, even greater precision is already available.'

When units of length and time are both based on the oscillations of an atom, the question of whether the speed of light is constant becomes trivial. For example, if the unit of length were defined as that of m wavelengths of light in a certain atomic radiation, and the unit of time as n periods of the same oscillations, then in unit time this radiation would travel a distance of n/m units. In other words, the radiation is automatically propagated at the speed n/m, the ratio of two chosen numbers. Einstein's postulate, however, was not trivial because it was originally made with regard to distances as measured by rods. We note, too, that in experiments of the Michelson–Morley and Kennedy–Thorndike type, the distances between parts of the apparatus are maintained, in effect, by rods.

Many writers (see, for example, Arzèlies [2]) regard *natural time,* which is based on the notion of constant light velocity over an outward and return journey, as the most satisfactory *theoretical* time standard in relativity. Natural time is that measured by an ideal clock known as the Einstein–Langevin clock, which consists of a pair of parallel mirrors (*A, B* in Fig. 2) attached to a rigid rod. (By rigid we mean of constant length according to a pre-chosen length standard.) A light ray travels to and fro between the mirrors, being successively reflected at *A* and *B*, and natural time is metered by successive reflections at *A*. No time measurement at a distance is involved. However, natural time at the point *A* is not uniquely defined without the introduction of certain assumptions. For example, do two identical Einstein–Langevin clocks *AB, AB'* set at an angle, as in Figure 3, agree in rate? The assumption that space is *isotropic,* i.e. that the properties of space are the same in all directions at any point, is made in relativity theory on the basis of experience. If space is isotropic, the clocks *will* agree. Isotropy is suggested by the Michelson–Morley experiment (c.f. the remarks of H. P. Robertson in §3), the apparatus being not unlike a pair of Einstein–Langevin clocks.

Whether natural time is in precise agreement with the atomic time standard is a matter for experiment, rather than theory, to decide.

5. Simultaneity and the synchronization of clocks

What is meant by simultaneity? Two occurrences, or *events*, are said to be simultaneous if they occur at the same time, the time of each event being made by reference to a nearby clock. When the two events occur at points which are close together, the same clock can be used in the timing and it is an easy matter to decide whether the events are simultaneous or not. But when the events are at widely separated points two clocks must be used, and these have to be synchronized beforehand according to an agreed criterion. In practice, we do not always have clocks near events in which we are interested (like the explosion of a distant star) and the test of simultaneity has to be made by a different, though theoretically equivalent, procedure. In Newtonian theory the whole problem is hidden by the assumption that 'time is absolute', which is essentially another way of saying 'all clocks can be kept synchronized'.

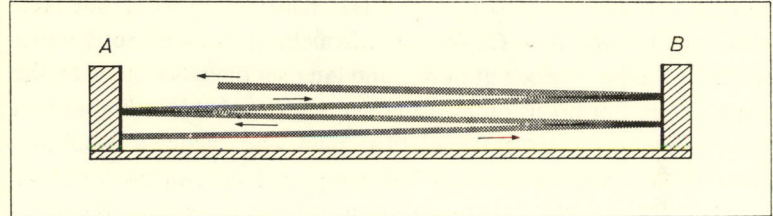

Fig. 2.   *The Einstein-Langevin clock*

It is important to notice here that two issues are involved in the synchronization of a clock *B* with a clock *A*, whatever criterion is used. First, it is necessary to adjust the reading of clock *B*, on one occasion, so that it is 'correct' immediately after adjustment. This is the familiar task of putting a clock right by resetting the hands, and is what we shall normally mean by synchronization. Secondly, it is

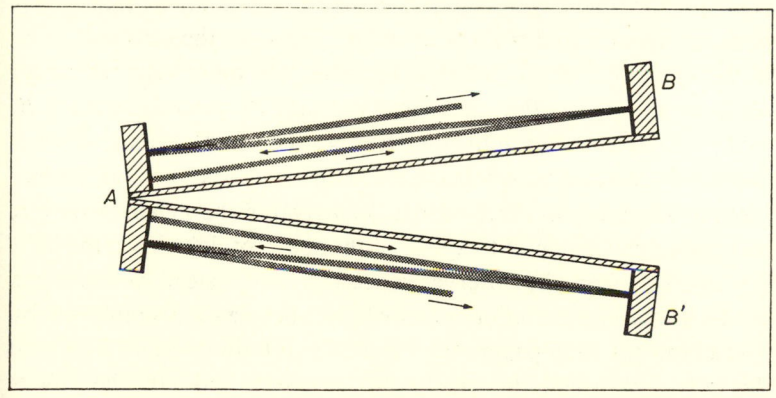

Fig. 3.   *A pair of identical Einstein-Langevin clocks inclined at an arbitrary angle*

41

necessary to regulate the mechanism of clock B so that it *remains* in synchronization with A. (Sometimes the word synchronization is used to denote both adjustment and regulation of a clock.)

Suppose that A and B are identical clocks, each measuring standard time, at rest in any inertial frame. One possible procedure for the synchronization of B with A is by means of a third standard clock C which is carried from A to B. First C is synchronized with A, and later B is synchronized with C. We use this method when we set a wrist-watch by a reliable clock at home, and later set the office clock by the wrist-watch. For precision measurements, a drawback is that motion might somehow affect clock C. (The man who carries a pendulum clock while his wrist-watch is being repaired encounters a similar problem.) Thus, the synchronization may depend on how C is transported from A to B.

To eliminate effects due to apparent forces on C, and to keep the state of motion of C as near as possible to that of A and B the transportation should be made extremely slowly and smoothly. Although impracticable in some circumstances, the slow transport method of synchronization is possible, in principle, in both Newtonian and relativity theories (Fig. 4a).

A second, and in many respects preferable, method of synchronizing clocks is by the sending of signals (Fig. 4b). Because no known type of signal is propagated faster than light, due allowance has to be made for the time of propagation. Suppose, for example, that a light or radio signal is sent from A at three o'clock, in the direction of B. If the distance from A to B is l metres, we intuitively expect an observer at A to say that the signal will take l/c seconds to reach B, where c is the speed of light in metres per second. Hence the clock B might be regarded as synchronized with the clock A if the former reads l/c seconds past three o'clock on arrival of the signal. However, it is not very meaningful to speak of the speed of light when the light is a long way from the observer concerned, and instead we make use of the following definition of clock synchronization employed by Einstein in his 1905 paper. (In Einstein's notation, A and B denote the *points* at which the clocks are situated, and not the clocks themselves.)

Fig. 4(a). *Clock synchronization by slow transport*
   (b). *Testing the Einstein synchronization of clocks*

'Let a ray of light start at the "A time" t_A from A towards B, let it at the "B time" t_B be reflected at B in the direction of A, and arrive again at A at the "A time" t'_A.

In accordance with the definition the two clocks synchronize if

$$t_B - t_A = t'_A - t_B.'\tag{2.1}$$

In other words, the clock at B is synchronized with the clock at A if

$$t_B = \tfrac{1}{2}(t_A + t'_A),\tag{2.2}$$

which is the mean of the 'A times' at which the light signal is sent and received back again at A.

The procedure can be pictured in a simple way. An observer stationed at A briefly illuminates B with a flash of light from his lamp. He reads the 'B time' at the moment of illumination. If this reading is the mean of his own times of lighting his lamp and of seeing the illuminated B clock, then the B clock is synchronized with the A clock (Fig. 4b).

By using a permanently synchronized distant clock, the observer at

A can allot a time to a distant event. This means that he can now assign an average speed to light over long distances, and the Einstein definition is such that the light from the A observer's lamp travels at the same speed on the outward and return journeys, because it takes equal times to travel each way. If the distance $AB = l$, then $t'_A = t_A + 2l/c$, and so the synchronization condition is that

$$t_B = \tfrac{1}{2}(t_A + t'_A) = t_A + l/c, \tag{2.3}$$

where c is the light speed in either direction. This agrees with an earlier intuitive suggestion.

By the assumption that the light speed is the same at every point of an inertial frame, we can show that Einstein clock synchronization is symmetrical for two observers. That is to say, it makes no difference whether the observer at A, or a corresponding observer at B, performs the test. This follows from the relation, obtained using (2.3),

$$t'_A = t_A + 2l/c = t_B + l/c, \tag{2.4}$$

which is the appropriate form of the criterion (2.3) for use by the observer at B.

To be useful, any criterion of clock synchronization must be applicable to all clocks at rest in a given inertial reference system. Einstein made the explicit assumption that 'the definition of synchronization is free from contradictions, and possible for any number of points'. We can, in fact, verify that this is the case. It is sufficient to show that if any third clock C is synchronized with B, then it is automatically synchronized, also, with A. Suppose that a light signal leaves A at time t_A, and travels along the triangular path (by means of instantaneous reflection at B and C) from A to B, from B to C, and finally back to A. The signal arrives at clocks B and C as they respectively read:

$$t_B = t_A + AB/c \qquad \text{(clock } B)$$
$$t_C = (t_A + AB/c) + BC/c \qquad \text{(clock } C)$$

because of the synchronization of B with A, and C with B.

Since the total distance travelled in the closed path is $AB + BC + CA$, the signal arrives back at A as this clock reads

$$\bar{t}_A = t_A + (AB + BC + CA)/c.$$

Therefore,

$$t_A = t_C + CA/c,$$

and so clocks C and A are synchronized.

We consider next the question of synchronizing a moving standard clock with one at rest in an inertial frame of reference. Let A denote a clock at rest in a particular inertial frame, denoted by S, and let B be a second clock, in any state of motion. We can synchronize clock B with clock A by means of a light signal sent from A and reflected back by B. It does not follow, however, that B will of itself *remain* synchronized with A. Nor does it follow that an observer moving with B would agree that the two clocks had been synchronized in the first place.

At this stage we can say little about an observer with B's point of view, unless we restrict B's motion to be inertial, because a deeper analysis is needed before one can discuss observations by accelerated observers in special relativity theory. Therefore, we shall suppose that B is at rest in a second inertial frame. By a simple argument, Einstein deduced from the hypothesis of the constancy of the velocity of light that two-way synchronization of clocks A and B is impossible. This is because observers in uniform relative motion do not agree on the simultaneity of separated events, whereas clock synchronization is simply a procedure for setting clocks to agree at 'the same time'.

Consider a train travelling steadily along a straight track. At the front and rear ends, denoted by M' and N', are carried signal lamps. The lamp at the front end, M', flashes as it passes a point M on the embankment, and similarly, the lamp at the rear end, N', flashes as it passes a point N on the embankment (Fig. 5a). If the points M and N are suitably spaced, the flashes will occur simultaneously according to an observer seated on the embankment, and if the observer is at the mid-point O of MN, the light rays from the lamps will reach him simultaneously.

Next consider a second observer, who is travelling on the train. He is seated in the central compartment, at the very mid-point O' of the train. His own criterion for the light rays to be emitted simultaneously is that they reach O' simultaneously. By symmetry, the observer on the embankment reckons that O' passes O at the same time as M' passes M and N' passes N. Therefore, by the time the rays arrive at O, which they do simultaneously, the point O' will have passed O. Thus, the rays do not arrive simultaneously at O' (Fig. 5b) and so the observer on the train does not agree with the observer on the embankment that they were *emitted* simultaneously.

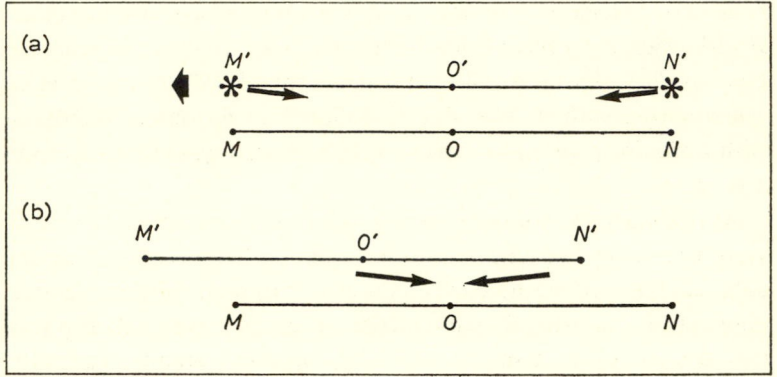

Fig. 5.

It should be borne in mind that the simultaneous receipt of light signals emitted at two events is a criterion of the simultaneity of the events themselves only when the observer concerned is at the same distance from each event.

Differences in simultaneity criteria are quite negligible at train speeds, but not at the high speeds of possible future space-travel. In the latter case, clocks in the frames of the earth and of a space-ship may separately be synchronized, but any clock in the one frame cannot be synchronized with all clocks in the other frame to the mutual satisfaction of earthmen and space-travellers.

The arbitrary nature of Einstein's definition of synchronization has frequently caused objections to clock paradox arguments which involve 'time-at-a-distance'. (See, for example, Dingle [56].) Such objections are unwarranted if time-at-a-distance is used consistently in a calculation. It serves merely as a coordinate, in the same way as other (equally arbitrary) coordinates, as a disposable aid in making unambiguous predictions from physical laws.

6. The Lorentz transformation

Once it is accepted that simultaneity is a relative rather than an absolute property of a pair of events, i.e. is defined only relative to the chosen reference frame, it becomes clear that other space and time measurements may depend, in an unfamiliar way, on the frame in which they are made. The relations between measurements made in different frames therefore have to be re-examined. Such a re-examination was initiated by Einstein, who first determined from his two basic postulates of relativity theory how the coordinates and time of a single event referred to one frame were related to those of the same event referred to a second frame.

We need to be a little more precise about the meaning of the term 'event'. An event is an actual or imagined occurrence so localized in space, and of such brief duration, that it can be regarded as occupying just one point in space and one instant in time. (Even an occurrence like the explosion of a star may be thought of as point-like and instantaneous in appropriate circumstances, e.g. when one is considering a large region of the universe over a long period of time.)

We also need to make explicit a previously implicit assumption that in every inertial reference system the geometry of space is Euclidean. In other words, if the right standards of distance, 'straightness', etc., are used, then we should be able to exhibit Euclidean theorems in physical space by building triangles out of rods, and so forth. (The assumption *could* be false, and indeed *is* false according to the general theory of relativity. But the deviations from Euclidean geometry, as predicted by general relativity, are significant only in the presence of extremely large gravitational fields or when regions of cosmological proportions are considered.) We further assume, on the basis of experience, that the 'right' distance standard is approximately the one we have chosen to use, whether this be the krypton wavelength or the platinum-iridium standard. The notion of 'straightness' is already incorporated in the concept of inertial motion. Well-isolated particles of matter (and light rays) move along straight line paths in terms of our standards; if these were the 'wrong' standards, they would not have found their present privileged place in scientific theory and measurement. Therefore, we may suppose that cartesian co-ordinates (x, y, z) can be used in any inertial system, the coordinates being directly related to distance measurements in the usual way. For example, if two particles are at rest at the points (x_1, y_1, z_1) and (x_2, y_2, z_2), then the measured distance between them is

$$\sqrt{[(x_1 - x_2)^2 + (y_1 - y_2)^2 + (z_1 - z_2)^2]},$$

in accordance with Pythagoras's theorem.

So much for preliminaries. Now, consider two inertial frames S and S' in which cartesian coordinates (x, y, z) and (x', y', z') are used. For simplicity, the corresponding coordinate axes in the two systems are to be parallel, and the x and x' axes are to lie along the same line, in the direction of the relative velocity of S' to S (Fig. 6).

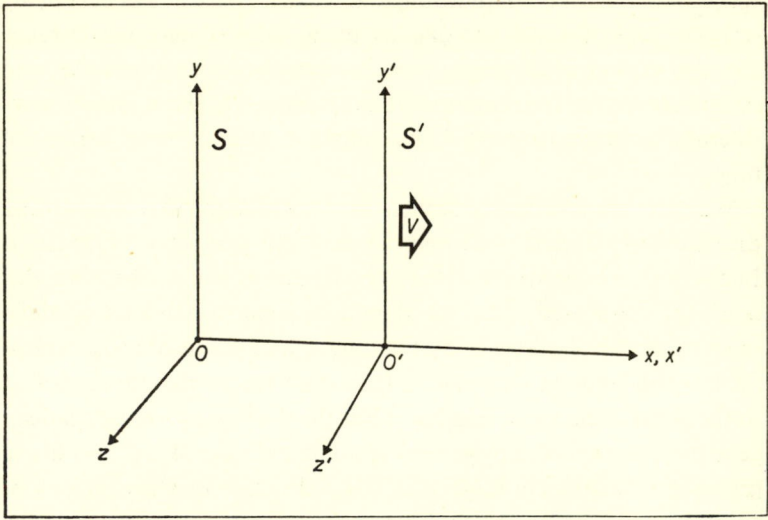

Fig. 6. *Inertial reference frames in 'standard configuration'*
O' coincides with O at $t' = t = 0$

The relative speed of the frames is V in the sense indicated in the Figure.

In the frame S time may be considered absolute, and the time of any event may be determined by reference to a nearby standard clock (synchronized with all other clocks in S), or by an equivalent method using a 'master' clock and light signals between the master clock and the event. We use the symbol t to denote time in this frame. Similarly time in the frame S' may be considered absolute, and the time of an event is determined in S' by an exactly similar procedure to that used in S. Time in S' is denoted by t'. To make full use of all possible simplifying assumptions, we suppose further that in each frame time is measured from the instant the origin O' in S' passes the origin O in S. In other words, this particular event occurs at $t = t' = 0$. None of our simplifying assumptions is restrictive in any physical sense, because we can orient coordinate axes and adjust clocks in *any* two inertial systems so that they are in this *standard configuration* with each other.

The problem to be considered is this. How are the space and time coordinates x, y, z and t, of any event, related to the coordinates x', y', z' and t' of the same event? First, let us write down for comparison the corresponding pre-relativistic formulae. According to Newtonian principles we have $t' = t$. Also, $OO' = Vt$, and so the required formulae are:

$$x = x' + Vt,$$
$$y = y',$$
$$z = z',$$
$$t = t',$$

(2.5)

or equivalently,

$$x' = x - Vt,$$
$$y' = y,$$
$$z' = z,$$
$$t' = t.$$

(2.6)

Equations (2.6) are known as the *Galilean transformation*, after Galileo Galilei (1564–1642).

The Galilean transformation is, of course, incompatible with the relativistic postulates of the constancy of the velocity of light. We do not need to work through the train example on simultaneity to see this. For, the classical law of 'addition of velocities' is inconsistent with the existence of a light velocity which is the same in all inertial frames. Suppose that a light ray along the x axis is emitted from O at

$t = 0$. At any later time, $t = t_1$ say, the ray will have reached the point $x = ct_1$ ($y = 0$, $z = 0$). The arrival of the ray at this point is an event (e.g. the illumination of a dust particle which happens to be located there), whose coordinates are:

$$x = ct_1, y = 0, z = 0, t = t_1.$$

This event in S' has the coordinates, according to (2.6),

$$x' = (c-V)t_1, y' = 0, z' = 0, t' = t_1,$$

and so we get $x' = (c-V)t'$. Therefore, the light speed in S' is $c - V$, and not c.

Einstein arrived at a new transformation to replace the Galilean transformation of space and time coordinates by appeal to the two principles referred to, namely the principle of relativity and the principle of the constancy of the velocity of light. These were expressed explicitly in his 1905 paper in the following terms:

'1. The laws by which the states of physical systems undergo change are not affected, whether these changes of state be referred to the one or the other of two systems of coordinates in uniform translatory motion.

2. Any ray of light moves in the 'stationary' system of coordinates with the determined velocity c, whether the ray be emitted by a stationary or by a moving body. Hence

$$\text{velocity} = \frac{\text{light path}}{\text{time interval}},$$

where time interval is to be taken in the sense of the definition in §1 [i.e. using synchronized standard clocks].'

A 'stationary' system was to mean any chosen inertial system. From 1, we can infer that well-isolated free particles move uniformly in both frames S and S', since these are the inertial motions. This fact, together with minor additional assumptions which need not concern us here, leads to the conclusion that the relativistic transformation equations, like those of the Galilean transformation, must be *linear* equations. That is to say, each of the quantities x', y', z', t' must be equal to an expression of the form

$$a_0 + a_1 x + a_2 y + a_3 z + a_4 t,$$

where the a's are all constants, possibly depending on V. The proof is mathematical, and is to be found in most standard text-books on relativity theory. The right-hand sides in (2.6) are all simple cases of such linear expressions.

From 2, since either system might be called 'stationary', it follows that a ray of light emitted from O at the instant $t = 0$ will have reached a distance ct from O, measured in S, at any time t. Therefore, whatever the direction in which the ray travels, the front of the ray will have reached some point (x, y, z), at time t, such that $\sqrt{(x^2+y^2+z^2)} = ct$, or

$$x^2+y^2+z^2-c^2t^2 = 0. \tag{2.7}$$

Likewise, at any time t', the ray will have reached a certain point (x', y', z') in S' for which

$$x'^2+y'^2+z'^2-c^2t'^2 = 0, \tag{2.8}$$

because it started from O' at $t' = 0$. The two sets of events (2.7) and (2.8) are the same; they are the events at the front of all possible light rays sent out from a particular point in space at a definite instant. Therefore, the new transformation must be such that whenever x, y, z, t satisfy (2.7), then x', y', z', t' satisfy (2.8), and conversely. Very few linear transformations have this property, and for reference frames oriented as in Figure 6, only *one* proves acceptable. (Others have such failings as interchanging future and past in the two systems, etc.) The acceptable one is the *Lorentz transformation.**

$$x' = \frac{x-Vt}{\sqrt{(1-V^2/c^2)}}$$

$$y' = y,$$

$$z' = z, \tag{2.9}$$

$$t' = \frac{t-Vxc/^2}{\sqrt{(1-V^2/c^2)}}.$$

*Further details of the derivation are given in the author's book, *An Introduction to Relativity* [278]. The Lorentz transformation may also be derived from a variety of alternative postulates. Several references are to be found in Arzeliès' book [2].

(These equations are precisely those introduced by Lorentz in 1904 to relate the coordinates of events in two inertial systems, one of which was to be the ether system. The transformation continues to bear his name.)

Equations (2.9) can apply only if the magnitude of V is less than that of c, for otherwise the denominators on the right are zero or imaginary. Thus, the special theory of relativity denies the possibility that any observer and his frame can travel relative to a second observer and frame at speeds equal to or greater than c. According to further developments of the theory, no particle or energy can travel faster than light, either. We recall here Poincaré's reference in 1904 to 'a whole new mechanics, that we only succeed in catching a glimpse of, where inertia increasing with the velocity, the velocity of light would become an impassable limit'. The velocity c can, however, be exceeded when the transmission of mass (energy) or information is not involved. For example, if a beam of light from a lamp falls on part of a screen and the lamp is swung rapidly, the illuminated region may, in principle, travel across the screen with any finite velocity. Such a device cannot be used to send a signal from one part of the screen to another, although it can, of course, be used to send signals from the lamp to the vicinity of the screen. These signals travel with just the speed c.

The existence of a limiting speed is intimately bound up with the question of *causality*. The time ordering of two events need not be the same in two different reference systems. In the train example (§5), the simultaneous lighting of lamps according to the observer on the embankment is not simultaneous to the observer on the train. According to the travelling observer O', the lamp at the front end M' of the train lights first. If another observer, O'', were in a second passing train travelling in the opposite direction, then O'' would find that the lamp at the *rear* end N' was lit first (by a simple symmetry argument). Thus, the events of lamp-lighting can be simultaneous or can occur in either order, depending on the frame of the observer. There is no contradiction here. But events which are *causally* linked have a unique time-ordering. We know, for instance, that a distant telephone is answered only *after* we dial, and that there is no chance of observing a certain chemical reaction before the requisite chemicals are brought together. Physical laws would take a strange form in any frame where such statements were not true, and the principle of relativity tells us that these strange laws would apply in *all* frames, including our own. It will be clear later that if a moving point would need to travel at a speed greater than c to get from an event E_1 to an event E_2, then E_2 cannot be found by *all* observers to occur after E_1. In this case E_1 could not be a cause of E_2.

On the other hand, if the journey between the events can be made at speed less than c, then the time-ordering of the events is absolute, and E_1 could, conceivably, cause E_2.

Bizarre paradoxical situations can be contemplated in a relativistic world in which, however, signal speeds in excess of c are permitted. It would, for example, be possible to observe an event and then to take action to prevent that same event from occurring. David Bohm discusses such possibilities in *The Special Theory of Relativity* [246].

Note that when V is much smaller in magnitude than c, the denominator $\sqrt{(1 - V^2/c^2)}$ is nearly equal to 1, and the Lorentz transformation equations reduce approximately to those of the Galilean transformation. Relativistic effects are therefore negligible in most instances when low speeds are involved. Indeed, if relativistic effects were more prominent at low speeds, the special theory would have come into being sooner than it did.

The transformation expressing coordinates in S in terms of those in S' must be obtainable from (2.9) by replacing V by $-V$, and moving the primes over to the right-hand side. We get

$$
\begin{aligned}
x &= \frac{x' + Vt'}{\sqrt{(1 - V^2/c^2)}}, \\
y &= y', \\
z &= z', \\
t &= \frac{t' + Vx'/c^2}{\sqrt{(1 - V^2/c^2)}}.
\end{aligned}
\tag{2.10}
$$

This follows since the frame S moves relatively to S' with velocity $-V$ in the x' direction. Equations (2.10) constitute the *inverse transformation* to (2.9), and can alternatively be obtained from (2.9) by rearrangement.

Let us consider now the relativistic 'addition of velocities'. To take just a simple case, suppose that a body moves in the x' direction in S' with speed v, starting from O' at $t' = 0$. The subsequent position at any time t' is $x' = vt'$, with y' and z' constant. By (2.10), the position and time in S is

$$x = \frac{x'+Vt'}{\sqrt{(1-V^2/c^2)}} = \frac{(v+V)t'}{\sqrt{(1-V^2/c^2)}},$$

$$t = \frac{t'+Vx'/c^2}{\sqrt{(1-V^2/c^2)}} = \frac{(1+Vv/c^2)t'}{\sqrt{(1-V^2/c^2)}},$$

(2.11)

with y and z having the same constant values as y' and z', respectively. Therefore, if u denotes x/t,

$$u = \frac{v+V}{1+Vv/c^2},$$

(2.12)

which is the constant speed of the body in the x direction in S. Equation (2.12) replaces the pre-relativistic law $u = v+V$ for 'adding' velocities in a straight line.

When v and V are both positive, u is *less* than the sum of v and V. Some such relativistic result is to be expected, because however close V and v are to c, we know that u must not exceed c. For example, if $v = \frac{3}{4}c$, and $V = \frac{3}{4}c$, then on the basis of classical theory $u = 3c/2$, while according to relativity theory we get, by (2.12), $u = 24c/25$, which is less than c.

Velocities in other directions, not parallel to the relative velocity of the two inertial frames, may be treated similarly. It is found (surprisingly at first) that components of velocity in the y' and z' directions are different from those in the y and z directions, despite the fact that $y = y'$ and $z = z'$. This is because time-intervals are different in the two frames.

7. *Moving rulers and clocks*

It will be recalled that Fitzgerald and Lorentz suggested that bodies moving through the ether with speed v are contracted by the factor $\sqrt{(1-v^2/c^2)}$ in the direction of their motion. In relativity theory, this suggestion is of course meaningless because of the total absence of ether, but since the coordinates of events change in a non-classical way as we go from frame to frame the dimensions of a body might be expected to differ in some respect in different frames. The intimate relationship between coordinates and measurements makes it easy to test whether this is so.

Consider a metre stick AB which lies at rest parallel to the x' direction in the S' system, as in Fig. 7. In the S system, which is related to S' in the manner described in the last section, the rod moves parallel to itself with speed V. What is the measured length of

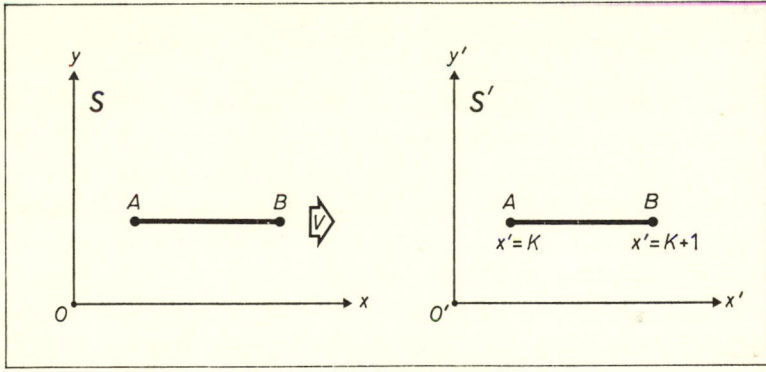

Fig. 7. *The metre stick AB is at rest in S'*

the stick in S? First, we have to make clear what the question means, for the length of an object in motion is a matter for definition. Even in principle we cannot place a metre stick at rest alongside the moving one and simply compare the two without making a conscious choice of how the comparison is to be made. In some sense, the length in S should be a measure of the distance between the ends A and B, if the word length is to retain any of its familiar meaning, but the times of observation of A and B still need to be specified. Guided by the classical concept of the length of a moving body, we shall assert that the observations of A and B are to be made *simultaneously* in the frame of measurement, and therefore that the length in S is the distance between the positions of the end points at a fixed value of t.

In the system S', the x' coordinates of A, B are constants which we

55

can write as $x'_A = K$, $x'_B = K+1$, for a certain number K. Denote the x coordinates of the ends in S by x_A, x_B, where each depends on the time t. The remaining space coordinates are the same in the two frames, and need not be considered further. At time t, we have by (2.9),

$$x'_A = K = \gamma(x_A - Vt),$$
$$x'_B = K+1 = \gamma(x_B - Vt),$$

where γ is used to denote $1/\sqrt{(1-V^2/c^2)}$. Therefore, by subtraction,

$$x'_B - x'_A = 1 = \gamma(x_B - x_A),$$

or

$$x_B - x_A = 1/\gamma = \sqrt{(1-V^2/c^2)}, \qquad \text{(metres)}$$

which is less than 1. *A moving rod is shorter than an identical one at rest.* Moreover, the contraction factor $\sqrt{(1-V^2/c^2)}$ is exactly the same as the Fitzgerald–Lorentz factor, although it now refers to the motion of a body relative to any inertial frame. The name Fitzgerald–Lorentz contraction is now commonly used in reference to the relativistic contraction phenomenon.

By the symmetrical relationship of the two systems S and S', it follows that a metre stick at rest in S appears shortened in S'. This may, of course, be verified from (2.10) directly. (Think of two witches on identical broomsticks. As they glide past each other, each notes with pride that her own status-symbol is the longer!) The result is only apparently contradictory, because the comparison of identical sticks is made according to different criteria in the two systems; in each system its own criterion of simultaneity is used. When a stick is at rest in a direction perpendicular to that of the relative motion of the frames, there is no contraction because y and z coordinates do not change in the Lorentz transformation (2.9). The length of a stick in its system of rest is called the *rest* length or *proper* length.

A perpetual question is 'Is the Fitzgerald–Lorentz contraction "real"?' Doubts usually arise because lengths of bodies in motion are arbitrary to the extent that the adoption of a particular definition of simultaneity (such as that due to Einstein, which we use) is arbitrary. But we can answer emphatically that the phenomenon is real inasmuch as the same length measurement procedure will give different results according to the classical (Newtonian) and relativity theories. This is clearly seen in the following example. Let two identical parallel rods AB, $A'B'$ move with speed V in opposite directions, so that they glide past each other, as indicated in Fig. 8. As A passes A', its position is noted in S by a nearby observer. Similarly, as B passes B', its position is noted in S by another convenient observer. The distance between the two positions is subsequently measured in S at leisure. The relativistic prediction for the measured distance is $\sqrt{(1 - V^2/c^2)}$ times the value according to classical theory.

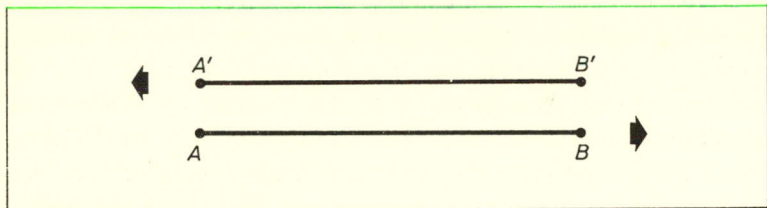

Fig. 8

Having dealt with the question of lengths in different systems, let us now consider the analogous problem involving time-intervals, which is the more important to this book. In particular, we need to determine the rate of a moving clock. Suppose that a standard clock at rest in S' ticks at 1 second intervals; what is the time-interval between the ticks in S?

In order to apply the Lorentz transformation to this problem, we need to associate events with the ticks. These events could be the emission of flashes of light from a lamp of some kind which is attached to the clock and governed by its mechanism. Or the events could simply be the showing by the dial of the readings 12 o'clock, 1 second past 12, 2 seconds past 12, etc. Either type of event fixes a position and an instant in the life of the clock. If the clock is located at the point (x', y', z') in the S' system, then two successive ticks are events with coordinates (x', y', z', T') and $(x', y', z', T'+1)$ for a certain number T'. According to (2.10), the times of the same events in S are given by

$$t_1 = \gamma(T' + Vx'/c^2), \qquad \text{(first tick)}$$
$$t_2 = \gamma(T' + 1 + Vx'/c^2), \qquad \text{(second tick)}$$

and so, on subtracting, we have

$$t_2 - t_1 = \gamma = \frac{1}{\sqrt{(1 - V^2/c^2)}} \quad \text{(seconds)}. \qquad (2.13)$$

This is the required time-interval in S. It is greater than one second, and therefore the moving clock appears to run slow.

By symmetry, it follows that a standard clock at rest in S appears in S' to run slow, as may also be verified directly from the transformation equations. As with the mutual contraction of lengths, there is no contradiction in this. In the one case, a particular S' clock is compared with a succession of nearby S clocks, all of which are synchronized by the Einstein procedure applied in S. In the other case, a particular S clock is compared with a succession of S' clocks, synchronized by the Einstein procedure in S'. *Moving clocks run slow.* This phenomenon is called *time-dilation* or *time-dilatation*.

It is rather difficult, at first, to visualize two systems of clocks, each of which runs slow according to the other. A simple example will serve to clarify the situation. In Fig. 9 is shown a formation of three space-ships travelling across the solar system. Their speed is assumed to be $\frac{1}{2}\sqrt{3}c$, or approximately $0.87c$, because for this special and rather high value of V, γ has the convenient value 2. The ships are equally spaced and go diametrically across the orbit of the outermost planet, Pluto, passing near the sun on the way. The spacing is approximately 5600 million km (a little less than the mean radius of Pluto's orbit) between successive ships, so that they pass the sun at intervals of six hours solar system time (SST) and each ship takes about twelve hours SST to make the crossing. Fig. 9 consists of three solar system 'snapshots', a snapshot being a picture of simultaneous events in a given frame. The snapshots, taken at six-hour intervals, show the time in hours indicated by three solar system clocks located at the beginning (A), mid-point (the sun B), and end (C) of the crossing. Also shown are the readings of synchronized clocks carried in the space-ships. Time zero, in each reference system, has been chosen arbitrarily as the instant when the first ship passes C.

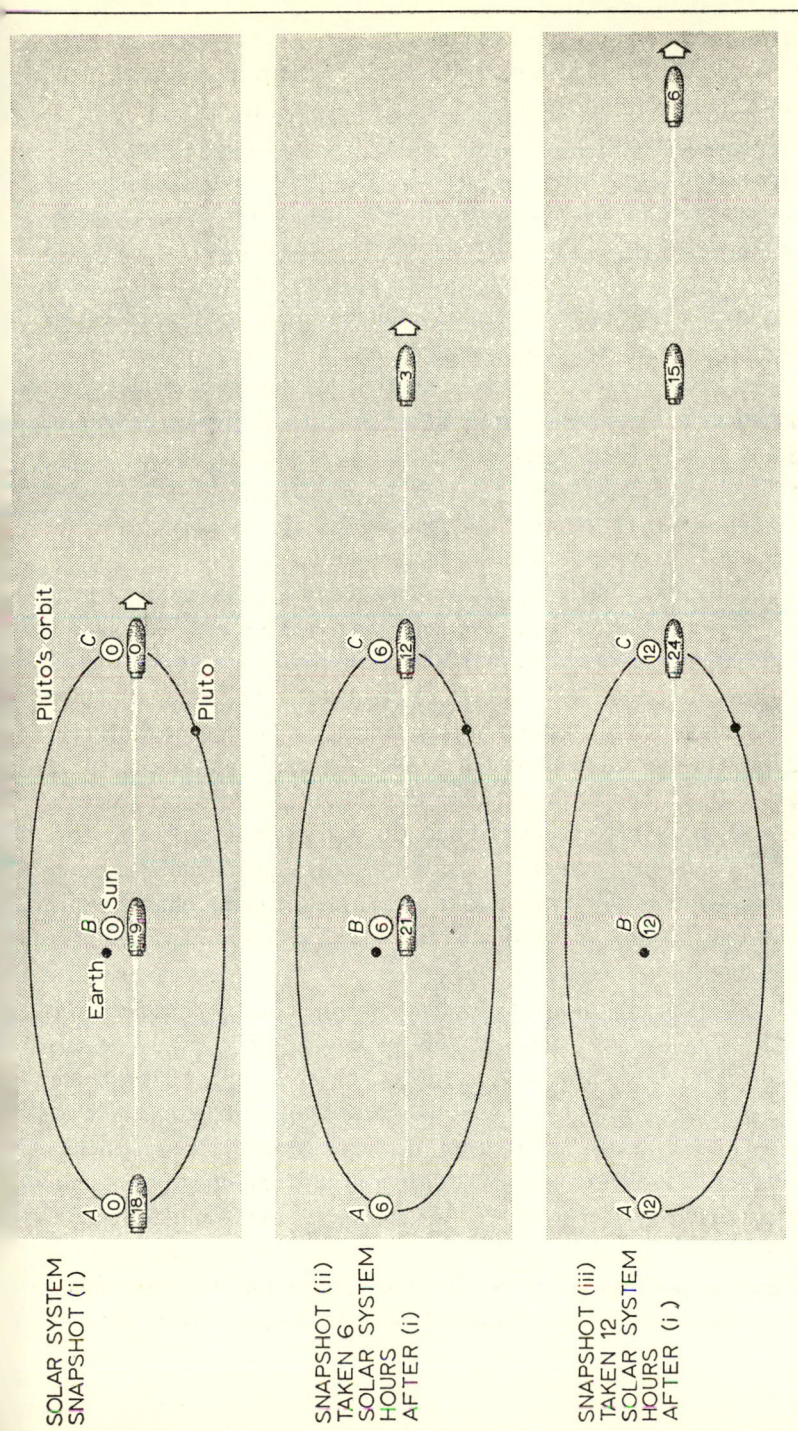

Fig. 9. *Three solar system snapshots of the crossing of a space-ship convoy, showing clock readings*

Notice that in snapshot (i), the clock in the rear ship is eighteen hours ahead of the adjacent solar system clock at A, but in (ii) it is only fifteen hours ahead of the adjacent one at B, and in (iii) it is only twelve hours ahead of the adjacent one at C. The ship's clock is thus running slow in SST. But notice also that the solar system clock at C agrees with that of the nearest space-ship clock in (i), is six hours behind the nearest one in (ii), and is twelve hours behind the nearest one in (iii). This demonstrates that solar-system clocks are likewise running slow in the ships' reference frame.

There is much experimental material relating to time-dilatation. We consider this in detail in Chapter 5, but it will perhaps be helpful at this stage to describe just one piece of evidence for the phenomenon. This concerns the lifetimes of μ mesons, which are a different kind of meson from the π mesons considered in §3. A variation of the lifetime with speed is attributed to time-dilatation.

The discovery of the μ meson occurred shortly before the Second World War, when various workers were making advances in the study of charged particles in cosmic radiation. It was found that at sea-level, cosmic radiation could be divided into two compartments, penetrating (or 'hard') and non-penetrating (or 'soft'), which occurred roughly in the proportions 2:1 or 3:1. The hard component would pass through considerable thicknesses of rock or penetrate deeply into lakes, whereas the soft component was stopped fairly easily by, say, a lead plate. After the war, the study of cosmic radiation intensified, with the wide use of a technique involving photographic emulsions exposed at various altitudes, and many more discoveries were made.

Investigations showed that primary cosmic radiation reaching the earth's atmosphere consists mainly of atomic nuclei, and that encounters by these nuclei with particles of air in the upper atmosphere leads to the production of π mesons. These charged π mesons quickly decay, usually with the emission of μ mesons and other particles, *neutrinos*. The μ mesons are the penetrating component of cosmic radiation observed at sea-level, the other component being formed of electrons.

When μ mesons have been produced, they too are short-lived. Each meson decays into an electron, which is easy to detect, and two neutrinos. *It is the average life of a μ meson from production to decay which concerns us here.* The *rest* lifetime can be determined by bringing the meson quickly to rest in an absorbing material and measuring the delay before the decay electron emerges. Currently, the accepted figure is about $2 \cdot 2 \times 10^{-6}$ seconds.

Lifetime measurements of high-speed μ mesons were made on many occasions in the 1940s. The procedure involved a count of the numbers present at high and low altitudes. There are three possibilities for mesons travelling downwards from a high altitude. They can (i) reach low altitude intact, (ii) decay on the way down, or (iii) be absorbed by air on the way down. If the proportion in (iii) is correctly estimated, the fraction of the total which decay before reaching low altitude is determined and so the mean lifetime can be calculated.

In 1940, B. Rossi, N. Hilberry and J. B. Hoag [293] described an experiment in which detecting apparatus was taken to various locations in Colorado at altitudes of up to 4300 metres. The sites chosen were at Denver (1616 m), Echo Lake (3240 m) and the top of Mt Evans (4300 m). Table 1 shows how the counts depended on

TABLE 1

Location	Carbon absorber?	Counts per minute (corrected average)
Mt Evans	No	$11 \cdot 79 \pm 0 \cdot 070$
(4300 m)	Yes	$10 \cdot 76 \pm 0 \cdot 114$
Echo Lake	No	$9 \cdot 65 \pm 0 \cdot 046$
(3240 m)	Yes	$8 \cdot 72 \pm 0 \cdot 097$
Denver	No	$6 \cdot 84 \pm 0 \cdot 039$
(1616 m)	Yes	$6 \cdot 36 \pm 0 \cdot 079$

altitude. Counts were made with and without a carbon absorber placed in front of the apparatus, the absorber being designed to have the same facility for absorption as the deep layers of air between altitudes. The differences in the counts *with* an absorber, at different heights, are therefore due only to decay occurring on the way down.

The third column shows that about 60 per cent of mesons present at 4300 m survive intact while travelling a further 2700 m downwards. The figures are found to be quite in accord with the relativistic prediction of a much longer life for the μ mesons in high-speed flight compared with the life of those at rest. The meson acts as a clock measuring, on average, an interval of $2 \cdot 2 \times 10^{-6}$ s in its own frame. But in the earth frame, the clock runs slow by the factor $\sqrt{(1 - V^2/c^2)}$, where V is the speed of flight, and the lifetime of the particle is extended accordingly.

The same explanation applies to the simpler, qualitative, observation that a considerable number of μ mesons reach the earth's

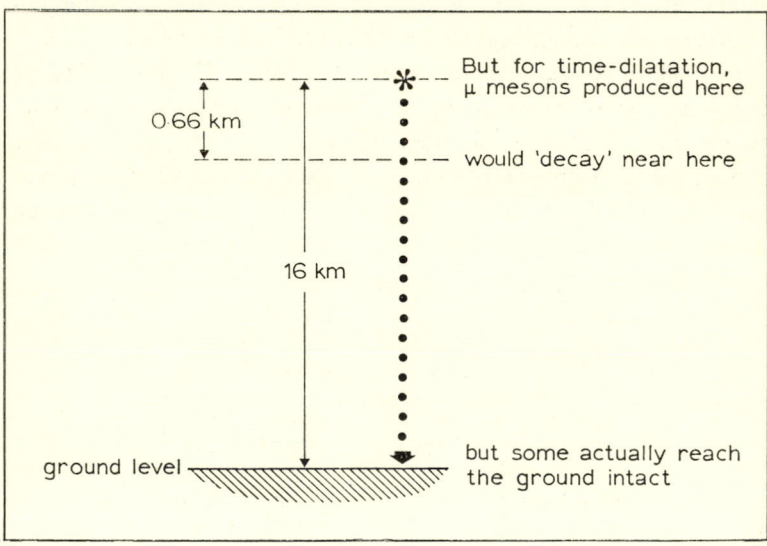

Fig. 10

surface. For, they are produced fairly high in the atmosphere, a typical height being 16 km. Even at speed c, the mesons would be able to travel a distance of only

$$2 \cdot 2 \times 10^{-6} \times 3 \times 10^5 = 0 \cdot 66 \text{ km}$$

towards the ground before decaying, if it were not for the relativistic effect (Fig. 10). Those that do reach the ground have speeds as high as 0·999c, or even higher, which means that γ may be as large as twenty or thirty, and that the meson can reasonably have survived a journey of twenty or thirty times 0·66 km since birth.

From the meson's point of view, its life-time is only $2·2 \times 10^{-6}$ s, but the distance from its birthplace to the ground is only $16/\gamma$ km in its own frame, because of the Fitzgerald–Lorentz contraction.

This widely accepted interpretation of the meson observations has been criticized by E. G. Cullwick in connection with the clock paradox. In his book *Electromagnetism and Relativity* [45] he writes:

'the meson life is not directly measured but is inferred from such considerations as the density and estimated rate of production of mesons at different altitudes. In reading the literature of the subject (e.g. Rossi, Hilberry and Hoag), however, it becomes clear that the alleged reality of the relativistic "time dilatation" is not questioned, so that the interpretation of the data cannot be accepted as being entirely free from preconception.'

His own explanation of the ability of the μ meson to reach the earth's surface is:

'The primary cosmic particles, however, can have a clocked velocity greater than c since they do not originate in the reference system of the Earth. If, then, a meson is born with the velocity, or even a fraction of the velocity, of its cosmic parent it can also have a clocked velocity greater than c. It would then travel, in its true life, much farther in the Earth system than would be possible with a clocked velocity limited to c.'

But Cullwick's discussion of 'clocked velocities' in his book is confused and misleading, and his own interpretation of the meson's behaviour appears to have received little support elsewhere.

8. *Minkowski's four-dimensional space-time*

'The views of space and time which I wish to lay before you have sprung from the soil of experimental physics, and therein lies their strength. They are radical. Henceforth space by itself, and time by itself, are doomed to fade into mere shadows, and only a kind of union of the two will preserve an independent reality'.

So began an address by the mathematician Hermann Minkowski [285], at the 80th Assembly of German Natural Scientists and Physicians at Cologne, on September 21, 1908. The idea of representing any event (x, y, z, t) as a point in a four-dimensional space had earlier been considered by Poincaré, but the exploitation of the geometrical properties which could be associated with such a space was due to Minkowski. Minkowski's aim was to provide a geometrical basis for the transformation formulae of relativity in order to exhibit their underlying mathematical structure. His four-dimensional space came to be known as *space-time,* and the graphical representation of the motion of objects in space-time as a *space-time diagram.*

Let us begin to study Minkowski's approach by considering an object which moves along the x axis of an inertial system. To fix our ideas, suppose that the object is a rocket fired vertically from earth (the x axis being directed vertically upwards) and that we wish to present graphically the height attained in terms of (earth) time elapsed since firing. A common way of portraying this type of motion is by plotting the height x of the rocket against the elapsed time t in a rectangular graph. Axes Oxt are taken, where the origin O is the event of firing, and the whole motion is represented as a continuous line beginning at O.

For example, in Fig. 11 is shown the motion of a two-stage rocket. The first stage fell back to earth after being jettisoned. The second stage continued to accelerate away from earth until its fuel expired, when it began to decelerate under the gravitational pull of the earth (and the rest of the solar system) while heading for distant space.

Any event (x, t) which occurs on the x axis of a reference system can be represented by the point with coordinates (x, t) in a diagram of this sort. When the y and z coordinates are permanently zero, space-time is simply the two-dimensional space consisting of all possible points (x, t). Minkowski used the term *world* to denote space-time, and *world-point* to denote any point such as (x, t) in it. These terms are rather less used nowadays, although the expression *world-line* is universally used for the lines of motion of a (point-like) object in space-time.

When a particle moves in two spatial dimensions, as for example on a plane region of the surface of the earth, the space-time diagram of its motion is three-dimensional, as in Fig. 12. The plane of motion has here been taken as the xy plane, and rectangular cartesian axes

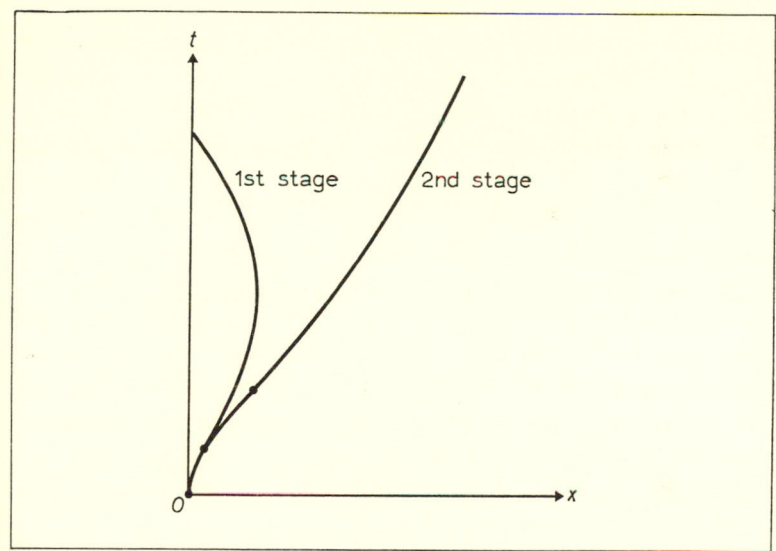

Fig. 11.   *World-lines of two stages of a rocket*

Oxyt introduced. The world-line of the particle is, in general, a twisted (i.e. non-plane) curve in this case. A body of non-negligible size can also be represented on the diagram; the world-lines of all its points occupy a tube, called a *world-tube*. The world-tube of a moving disc is shown in Fig. 12.

Finally, when motion takes place in all three space dimensions, space-time is four-dimensional and so cannot be fully treated graphically. It can, however, be described in full mathematical detail just as satisfactorily as in the other more restricted cases of one or two space dimensions. Space-time is therefore a mathematical entity in which time is treated on an equal footing with space, and not a mystical physical space in which one can 'move about' in four dimensions.

Much of the later discussion in this book concerns motion taking place in a straight line. It is therefore worthwhile investigating further properties of the xt type of space-time diagram.

The world-line of any particle at rest in the reference system considered is a straight line parallel to the time axis, Ot. The time

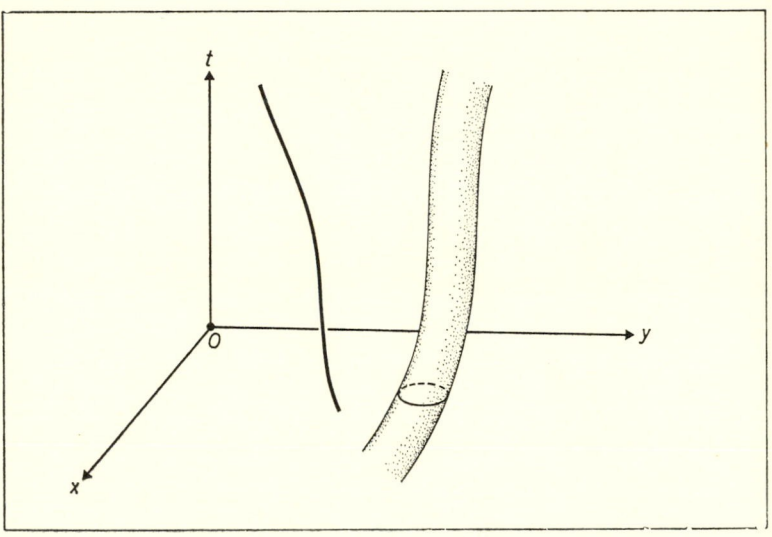

Fig. 12. *World-line of a particle and world-tube of a disc, both moving in the xy plane*

axis itself is the world-line of a particle at rest at $x = 0$. On the other hand, the x axis, and lines parallel to it, are not possible world-lines. For, they are the lines $t =$ constant which connect simultaneous events at different points, and no particle can be at two different points simultaneously. These lines are called *simultaneity lines*.

World-lines are never inclined at more than a certain angle with Ot, the limiting angle being that between Ot and the so-called *light-lines* (the lines representing the motion of light flashes), because the speed of a particle is always less than c. Imagine a lightflash to occur at the point $x = 0$ at time $t = 0$, so that the light is propagated along both the positive and negative x directions. The space-time path of light travelling in the positive x direction is the straight line $x = ct$, while that of light travelling in the opposite direction is the straight line $x = -ct$. These two lines, jointly given by the equation $x^2 - c^2 t^2 = 0$, divide the space-time into four regions as shown in Fig. 13. In fact, there is for present purposes no significant difference between the two

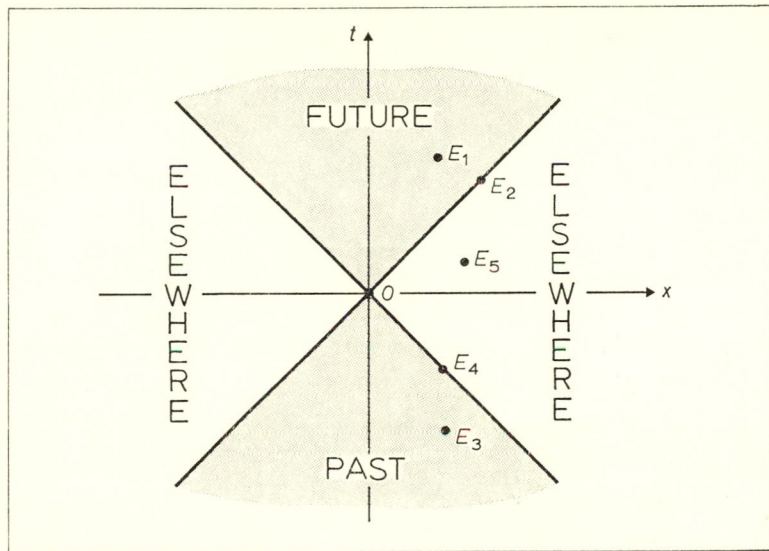

Fig. 13

regions marked 'elsewhere', and we regard the total number of regions as three. It will be shown that the division of space-time into these three regions has a simple but important interpretation and, moreover, the mode of division is one about which all inertial observers would agree.

To simplify the argument, let us choose units so as to make the velocity of light numerically equal to 1. This can be done, for example, by retaining the metre as the unit of distance and introducing $1/(3 \times 10^8)$ seconds (approximately) as the unit of time. If the space-time diagram is drawn with equal scales for x and t, then the light-lines through O (given by the equation $x^2 - t^2 = 0$) will both be inclined at 45° to Ot.

Let S denote the inertial reference system currently being used, and let S' be the system which moves along Ox with speed V (in the new units, so that $|V|$ is less than 1), the origins of the two systems coinciding at $t = t' = 0$ as usual. We shall first verify that the division of events is the same in the two systems, a 'future' event in Fig. 13 being also a 'future' event in the space-time diagram for S', and so on. Consider the expression of $x'^2 - t'^2$. By (2.9), with c put equal to 1, we have

$$x'^2 - t'^2 = \gamma^2(x - Vt)^2 - \gamma^2(t - Vx)^2$$
$$= \gamma^2(1 - V^2)(x^2 - t^2)$$
$$= x^2 - t^2,$$

since $\gamma^2 = 1/(1 - V^2)$. Therefore, the expression $x^2 - t^2$ does not change form (or is *invariant*) under the Lorentz transformation (2.9). But the division of events into regions is made essentially according to the sign of $x^2 - t^2$. The criteria for classifying an event or world-point (x, t) (which is not actually *on* one of the light-lines $x^2 - t^2 = 0$) as 'future', 'past' or 'elsewhere' are as follows:

Future: $x^2 - t^2$ negative; t positive,
Past: $x^2 - t^2$ negative; t negative,
Elsewhere: $x^2 - t^2$ positive.

Now, by (2.9),

$$t' = \gamma(t - Vx) = \gamma t(1 - Vx/t), \qquad (2.14)$$

and since $|x/t|$ is less than 1 for events in the 'future' or 'past' regions, it follows that the bracketed term is positive in (2.14) for these events, and hence that t' has the same sign as t. Therefore, the classification of events according to S':

Future: $x'^2 - t'^2$ negative; t' positive,
Past: $x'^2 - t'^2$ negative; t' negative,
Elsewhere: $x'^2 - t'^2$ positive,

is identical to that above, which relates to S.

The fundamental physical distinction between events in the various regions lies in their differing possibilities of communication with the event O. It is possible to make contact with any 'future' event E_1 by means of a signal sent from O with speed less than the light speed 1. Any event E_2 on the future (t positive) half of the light-lines can also be reached by a signal from O, the signal speed in this case being necessarily equal to 1. The event O can be reached from a 'past' event E_3, or an event E_4 on the past (t negative) half of the light-lines, by means of a signal with speed less than 1 in the former case, and speed equal to 1 in the latter.

Thus, O could cause or influence events such as E_1 or E_2, but not E_3 or E_4. Similarly, O could be caused by, or influenced by, events E_3 or E_4 but not E_1 or E_2.

Finally, an 'elsewhere' event E_5 can have no communication or causal relationship (in either direction) with O. For example, an observer present at the event O has no knowledge whatsoever *at that instant* of any 'elsewhere' event.

There is a somewhat different way of interpreting this division of events. Let the equation of the straight line OE_1 be $x = Ut$. Then, by the Lorentz transformation formulae it follows that both events O and E_1 occur at the point $x' = 0$ in the particular inertial system S' which has speed U relative to S. In S', the two events occur at the same place and so differ only in time. An analogous situation exists with regard to every 'past' event; there is always an inertial system in which the event occurs at the same place as O.

Events 'elsewhere' do not have this property, but there is always an observer who regards an event such as E_5 as simultaneous with O. In the frame of this observer the two events differ only in position, not in time, i.e., E_5 occurs *elsewhere* from O. To prove this, note that the straight line OE_5 has the equation $x = Kt$, for some constant K with magnitude greater than 1. Now, there exists a particular inertial frame S' whose speed relative to S is $1/K$, because $|1/K|$ is less than 1. By putting $x = Kt$, $V = 1/K$ in the Lorentz transformation formulae we get simply $t' = 0$. Therefore, all events on OE_5 occur simultaneously at $t' = 0$ in S' and in particular, the events O and E_5 themselves are simultaneous in S'.

The division of space-time into 'future', 'past' and 'elsewhere' regions may be made relative to any chosen event P, and not only relative to O. We simply draw the light-lines (inclined at 45° to the

t-axis) through the world-point P on the space-time diagram. This gives a quite different division of space-time, events in the three regions being related to P in the same way as those in the corresponding regions in Fig. 13 are related to O.

Furthermore, the full four-dimensional space-time may be divided into three regions in a similar way, although this cannot be faithfully represented graphically. However, in the intermediate case involving two space coordinates x, y and time t, a graphical description is possible. For example, the light-lines through the event (x_1, y_1, t_1) form the cone whose equation is

$$(x-x_1)^2 + (y-y_1)^2 = (t-t_1)^2.$$

This is illustrated in Fig. 14.

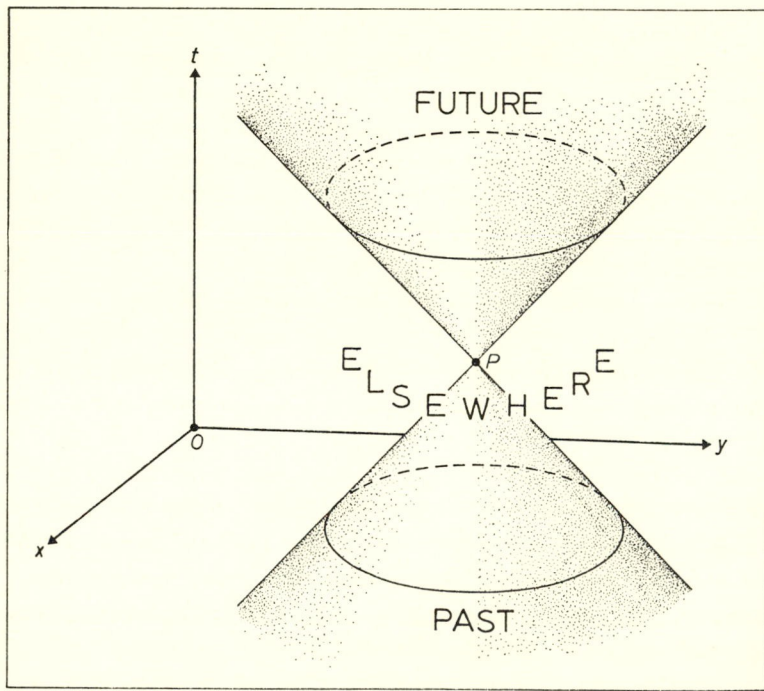

Fig. 14. *Future, past and elsewhere regions relative to the event* $P(x_1, y_1, t_1)$

For evident reasons, straight lines or displacements inclined at an angle of less than 45° to Ot are called *time-like*. Those inclined *at 45° to Ot are called *light-like*, and the remainder, *space-like*. The cone of light-lines in Fig. 14 is called the *light-cone* for the event P.

It is instructive to see how S' measurements may be made on the space-time diagram constructed for S. We shall work with just one space coordinate and the time, as before. The first step is to introduce

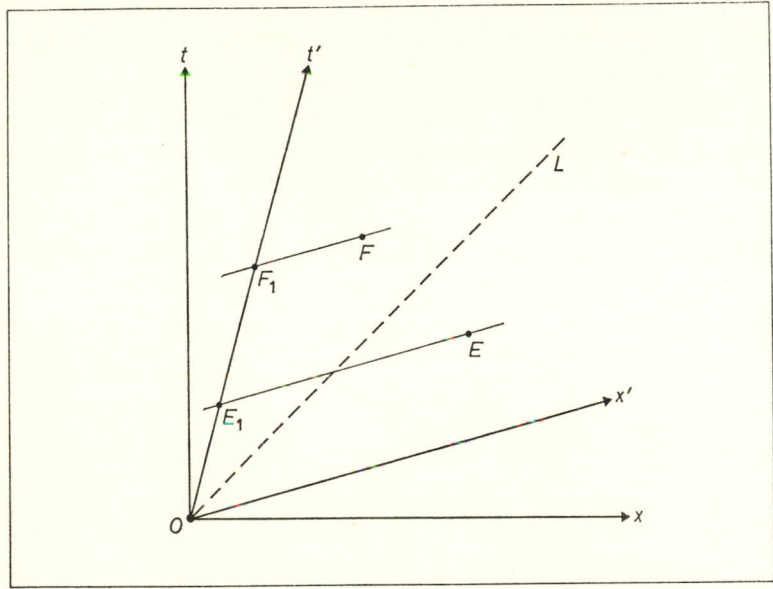

Fig. 15.   *The time interval in S' between the events E and F is equal to that between the events E_1 and F_1, and so can easily be measured when the correct scale has been determined*

the x' and t' axes. The t' axis is formed from the set of events $x' = 0$, i.e. $x = Vt$. Therefore, it appears as the straight line through O with slope V. Similarly, the x' axis is formed from the events $t' = 0$, and so by (2.14) it is the straight line $t = Vx$, as shown in Fig. 15. The two lines Ox' and Ot' are equally inclined to the lines Ox, Ot respectively, so that the light-line $OL(x = t)$ bisects both the angles xOt and $x'Ot'$. Lines parallel to Ox' are simultaneity lines in S', as they are given by equations of the form $x - Vt = $ constant, i.e. $t' = $ constant. Thus, the time-interval between any two events E, F will, according to S', be directly related to the distance on the diagram between the two S' simultaneity lines, one through each of the given events. The time-interval is most conveniently measured along Ot', between the intercepts E_1 and F_1 of the simultaneity lines, once the correct scale for the measurement has been determined.

It may be shown that unit time in S' is represented by a line of length

$$d = \sqrt{\frac{1+V^2}{1+V^2}},$$

(in the scale of the cartesian axes) measured parallel to Ot'. We must always divide by d to obtain the S' time-interval from the (oblique) ruler distance between simultaneity lines in the space-time diagram.

Length measurements in S' are performed in much the same way as time measurements, except that they are made parallel to Ox', not Ot'. The same scale factor d must be used in this case, too. The theory of the scale factor is omitted here for brevity, but is to be found in most books on relativity. (See, for example, Rosser [199].)

Chapter 3

THE PARADOX OF THE TWINS

'Time, young man, has taught us both a lesson!'

PLUTARCH, *Lives:* Themistocles, to Antiphales.

1. Paradox lost

We are now in a position to resolve the paradox of the twins described in Chapter 1. The following is perhaps the most popular of all 'conventional' arguments. To avoid, at this stage, the issue of whether a traveller's ageing is in accord with the standard clock that he carries, it is given in terms of clocks and not travellers.

As before, it is assumed that the earth's orbital motion around the sun, and its axial rotation (including various irregularities) have negligible effect on the rates of earthbound clocks. The rather loose but convenient phrase 'the inertial frame of the earth' is used to signify an inertial reference system in which the mean motion of these clocks is zero. The space-ship (we call it Nova) embarks on a long round-trip journey, the whole motion taking place along a straight line which is chosen to be the x axis of the earth's inertial frame, S, in which the earth is located at the origin O.

The journey envisaged consists of five stages. First, there is a period of acceleration, beginning with lift-off, when rocket motors are first used. At the end of this stage the ship's speed V along Ox is a signi-cant fraction of the speed of light. Secondly, there is a long period of uniform motion at the speed V. Thirdly, in the vicinity of a distant star or nebula the rocket motors are again used, in order to reverse the motion relative to earth. Fourthly, there is another long period of uniform motion at speed V, this time towards O. Fifthly, rocket motors are used for the last time to bring Nova back to rest on its arrival home. The whole journey is represented in a space-time diagram relating to the earth frame S, in Fig. 16, in which time $t = 0$ has been taken as the moment of lift-off.

In the Figure, the curved parts OL', $M'N'$ and $P'Q$ of the Nova

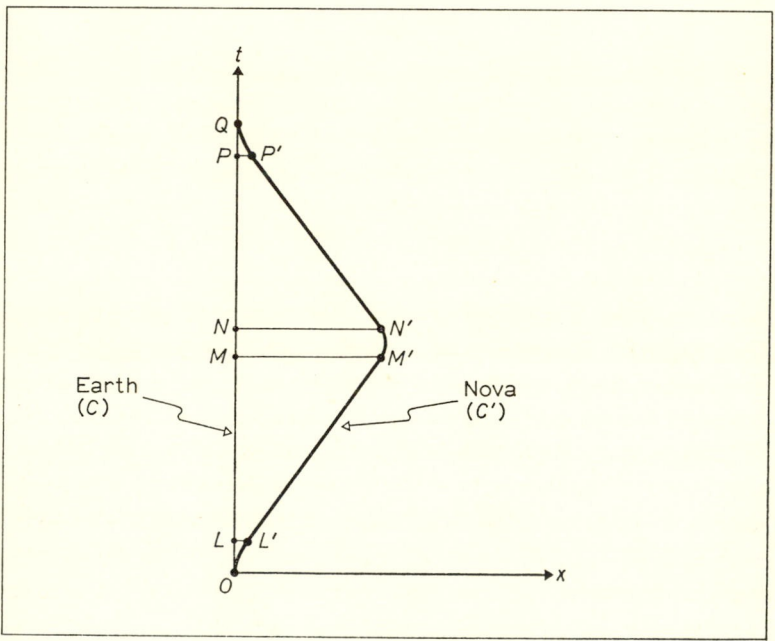

Fig. 16

world-line represents the three periods of acceleration, i.e. initial, mid-journey and terminal, when rocket motors are in use. Q is the event of final touch-down. The events L, M, N, P on the earth's world-line are those which are simultaneous with L', M', N', P', respectively, in S.

Let us first deal briefly with the periods of accelerated motion; the problem of acceleration will be discussed more fully in Chapter 4, §1. For present purposes, we can neglect the actual time spent in acceleration, as recorded both by an earth clock (denoted by C) and by a

Nova clock (denoted by C'), if the itinerary of the trip can be arranged so as to satisfy two conditions. These are:

(i) That the total C duration of the three periods of acceleration is a very small part of the C duration of the whole journey.

(ii) That the total duration of the acceleration periods as recorded by C' is at most comparable with that recorded by C (or by any of the synchronized clocks in S).

In connection with (ii), we note that clocks in daily use on earth are in reality in a constant state of acceleration, and so are those carried in present-day space-craft and aircraft. But these clocks are found to agree in rate with one another, and with the Newtonian time indicated by the motion of our planetary system, to within the accuracy of practicable measurements. It is therefore unimaginable that accelerations of magnitudes up to a few g can of themselves affect the rates of suitably constructed standard clocks by anything like, say, one-thousandth of one per cent. Therefore, if Nova's acceleration is not too great, but is sufficiently prolonged to attain the desired value for the speed V, condition (ii) will be satisfied.

There is no difficulty in complying with (i), since the total C duration of the uniform motion (represented by $LM+NP$) can be extended arbitrarily without affecting the C duration of the acceleration $(OL+MN+PR)$. It is only necessary that the visit be to a sufficiently distant star or nebula.

Now consider the uniform parts of the motion. The Nova clock C' records less time along $L'M'$ than C does along LM, because of time-dilatation. (The inequality is the right way round. In the earth system, C' goes slow and LM is the time taken for C' to record the interval $L'M'$, since L, L' and M, M' are pairs of simultaneous events.)

Similarly, the Nova clock C' records less time along the stretch of world-line $N'P'$ than the earth clock C records along the stretch NP.

Combining the last two results, we find that C' records less time than C over the whole journey, since the times spent in acceleration can be neglected.

The lack of symmetry in the motion of the two clocks is evident from the fact that there is no inertial system in which C' remains at rest. We cannot draw the world-line of C' as a straight line in any true space-time diagram. Rocket motors cause C' and not C to move non-uniformly, and so on. But the following argument, which makes no appeal to symmetry, is often put forward as paradoxical.

On the outward journey, C' is at rest in an inertial system, in which it must appear that C runs slow. On the return journey, C' is again at rest in an inertial system, in which again it must appear that C runs slow. On combining the two results it follows that, since the periods of acceleration are negligible, it is C and not C' that records the lesser total time. The contradiction with the previous result constitutes the paradox.

The fallacy in this argument is in the fact that not all the time recorded by C has been considered. It should be noted that the criterion of simultaneity in the first Nova inertial system is not the same as in the second. In Fig. 17, the simultaneity lines in the two

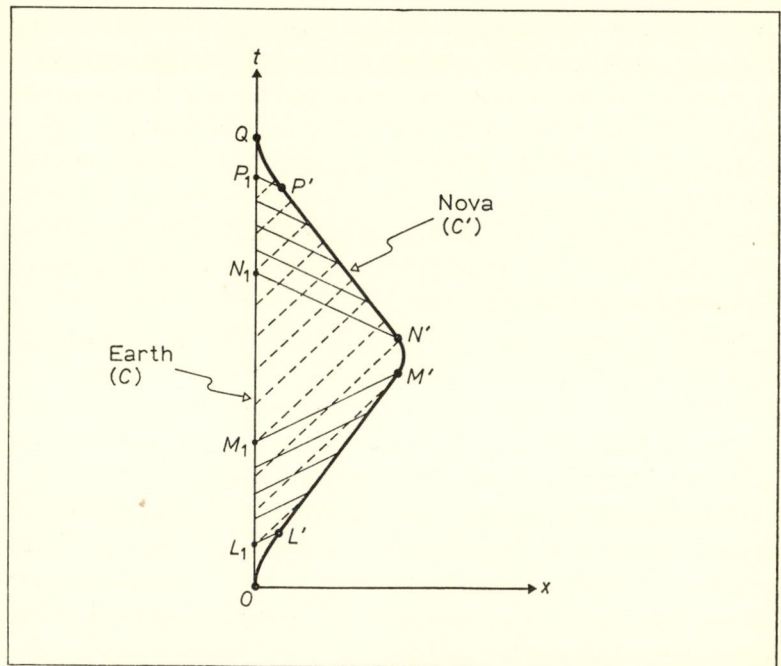

Fig. 17. *Full lines are some of the simultaneity lines in Nova's iner-tial systems. Broken lines represent light signals from earth to Nova*

Nova systems are shown (solid lines). On C's world-line, events L_1 and M_1 are simultaneous with L' and M', respectively, in the first Nova frame, and events N_1 and P_1 are simultaneous with N' and P' in the second. The slope of the simultaneity lines in the lower part of the diagram is V (see Chapter 2, §8; we are again putting $c = 1$), while those in the upper part of the diagram have slope $-V$, since this is Nova's speed along Ox on the return journey.

The C intervals represented by $L_1 M_1$ and $N_1 P_1$ are respectively less than the C' intervals represented by $L'M'$ and $N'P'$, as a consequence of time-dilatation. But the C time represented by $M_1 N_1$ has still to be included. The argument leading to the 'paradox' is therefore incomplete, and gives no indication of how this non-negligible period of C time is to be evaluated.

In Table 2, the correct deductions from the time-dilatation arguments from the earth and Nova viewpoints are summarized:

TABLE 2

Earth	*Nova*
$L'M'$ represents shorter time than LM $N'P'$ represents shorter time than NP	$L_1 M_1$ represents shorter time than $L'M'$ $N_1 P_1$ represents shorter time than $N'P'$
$OQ(C')$ represents shorter time than $OQ(C)$	$OM_1 + N_1 Q$ represents shorter time than $OQ(C')$

It is not to be concluded that there is a discontinuity in observations of earth made from Nova. Nor is it the case that earth clocks appear to run at an enormous rate during the ship's turn-round. What is observed from Nova through a telescope, for example, is propagated in the form of signals along light-lines (shown in Fig. 17), and these light-lines do not converge in any way on $M'N'$. The same is true of information received by radio. It is only the simultaneity criterion that undergoes a rapid change, and this has no physical significance since distant simultaneity has only a conventional meaning. However, it is perfectly in order to use a conventional concept like distant simultaneity as an aid in calculating an absolute result, such as the difference in trip times recorded by the separated and reunited clocks.

The 'out-and-out relativist', who believes simply that *all* motion is relative and that in the clock problem there is only the mutual recession and approach of two clocks, is of course denying the privileged role of the inertial reference frames and hence is denying special relativity. The special status of the inertial frames is a matter of experience, and likewise the special relativity theory will stand or fall in the light of experience. But the *logic* of the above argument, which concerns the heart of most clock paradox disputes, will remain totally unaffected by future observational evidence.

Some writers believe that only the general theory of relativity can deal adequately with the accelerated parts of the Nova motion. This viewpoint will be considered in Chapter 6. But even in the above argument based on special relativity, it is seen that although the actual times spent in accelerating are negligible, the existence of the accelerations is by no means negligible and must properly be taken into account.

The problem really involves the times recorded by clocks pursuing different *routes* in space-time from an event O to an event Q, and is a space-time analogue of a situation which occurs in everyday life. Imagine that Fig. 17 were part of a road map showing alternative routes from town O to town Q. The direct route is straight, and the other is straight except for bends of negligible length. To compare distances along the two routes we might choose to ignore the bends, but their existence would still show up in measurements of the straight stretches of road; we should not find the routes of nearly equal lengths. The analogue is really quite a close one, although in space-time the curved path represents the *shorter* time, because of the peculiar nature of space-time geometry.

W. Rindler [290] has interpreted the acceleration in a different way. He points out that in the brief period of initial acceleration Nova, in effect, covers a considerable part of the journey. Let the distance from earth to the farthest point of the journey be L (in the earth frame S). Before lift-off, therefore, L is the distance to be travelled. Immediately after the initial acceleration, Nova voyagers find that the distance to be travelled is only $L\sqrt{(1 - V^2)}$, because of Fitzgerald–

Lorentz contraction. Thus, during the brief period of acceleration, an effective distance of $L\{1 - \sqrt{(1 - V^2)}\}$ (which may be nearly equal to L) is covered. This of course is just another way in which the same final result may be interpreted.

2. We are all of us clocks

'We are all of us clocks whose faces tell the passing years.'

SIR ARTHUR EDDINGTON, *The Nature of the Physical World.*

To what extent is every living being a clock? Are man's ageing and his judgement of time-intervals in a fixed relation to the standard clock that he carries, independently of how he moves? Is the paradox of the twins identical to the paradox of the clocks? Of the many aspects of time-keeping in living beings, we shall touch briefly on only two. The first concerns man's conscious perception of time and his judgement of duration. The other relates to the unconscious measurement of time by means of biological or 'internal' clocks in animals (including man) and in plants.

Our conscious judgement of duration is notoriously unreliable; it is affected by many factors of a personal and environmental nature, by the way in which the limits of the judged interval are marked, the way the interval is filled, and so on. Furthermore, the findings of experimenters in this area are not always in agreement. Indeed, Herbert Woodrow in the *Handbook of Experimental Psychology* [305] puts the point rather more strongly:

'The data that have been accumulated in the illusive field of time perception show two outstanding characteristics. One is the conflicting nature of different experimenters; the other is the mentalistic nature of the data.'

Some uniform features do, however, emerge from the many investigations of experimental psychologists. (There is no dearth of material on the subject. The interested reader may, for example, commence a study by consulting just a few of the 567 articles in Fraisse's bibliography in *The Psychology of Time* [261]). Some of the most revealing information has come from the Swiss psychologist

Jean Piaget's studies of children.* Piaget became interested in the relationship between the perceptions of distance, speed and time. Since any one of these three quantities can be expressed mathematically in terms of the other two, the question arises as to whether the intuition of time is immediate or derived. This question had, in fact, been posed to Piaget by Einstein [288], who also asked whether the subjective intuition of time was 'integral with speed from the first'. The problem may be expressed symbolically thus: if d denotes distance, and t time (in the classical sense), is velocity v first encountered as a mentally calculated ratio $v = d/t$, or does the appreciation of velocity emerge at an earlier stage than the appreciation of time (the latter being derived from the equation $t = d/v$)? Or is it even the case that at first the three quantities t, d and v are largely uncorrelated?

From a number of simple but ingenious experiments, Piaget concluded that the child's conception of velocity is independent of that of time-interval. In the young child, velocity is appreciated from ordinal relations, with complete absence of measurability. Presented with two moving objects, the young child decides correctly which moves faster if overtaking occurs. Thus, it appreciates the ordering of positions, and changes in this ordering, but has no notion of speed estimation as would arise in repeated experiments involving only one moving figure. Appeal is made only to such concepts as 'behind', 'in front of', 'before' and 'after'.

One experiment which showed that the measurability of speed is not comprehended at this stage was extremely simple. A group of young children were shown two empty tunnels, and they agreed correctly that one tunnel was longer than the other (suggesting that the concept of length had been grasped). Then two dolls were caused to pass through the tunnels, one through each, so that the dolls were seen to enter together and to emerge together. The children asserted that the dolls travelled at the same speed (because they started and completed their respective journeys together). Thus, there was certainly no mental application of the equation $v = d/t$. When the

---

*For a bibliography of many of Piaget's writings, see John H. Flavell, *The Developmental Psychology of Jean Piaget* [260].

tunnels were removed, and the dolls moved as before, the children asserted that the doll travelling farther moved faster.

Piaget finds that as the child develops, it successively passes through a number of stages in this type of speed evaluation. In stage (i), only the final positions of moving figures is considered, the starting positions being left entirely out of account. In stage (ii), both starting and final positions are used in the assessment. Stage (iii) is one of refinement of the ordinal assessment ('superordinal'), when the changing distances between the figures is also taken into account. Finally, in stage (iv), the idea of measurability emerges.

In its early development, the child confuses successions of events in time with analogous successions in space. A movement from A to C via B can be correctly judged temporally (arrival at B occurs before arrival at C; time-interval from A to C exceeds time-interval from A to B, etc.) but the understanding of temporal order is only apparent. When several movements occur concurrently there is confusion. Each movement has its own 'local time' and equality of duration is not well understood. (One should perhaps be grateful that at this stage the child has not become indoctrinated with the hard-to-remove concept of 'absolute time'.) A dialogue with a six-year-old child during an experiment illustrates the type of confusion. Two figures, one yellow and one blue, were moved in the same direction along parallel straight paths on a table. The figures were started together but moved at different speeds, the yellow one going faster and farther but stopping first. 'Did they stop at the same time? *No, the yellow one stopped before the other.* Which one stopped first? *The blue one.* What do we do at mid-day? *We eat.* Now if we say that the yellow one stops at mid-day when does the blue one stop? (The race is re-run.) Does he stop at mid-day too, or before mid-day, or after mid-day? *Before midday.* Look! *Yes, the yellow one stops first. He moved for longer.* And the other one? *He stops before midday.*'

Piaget's experiments shed light on the way the adult perceives time, since many of the laws of ordinal and superordinal judgement are the same for adult and child. It appears that physical time is primarily evaluated by means of relations like $t = d/v$, but that in the early stages of measurability the child frequently neglects one of the two quantities on the right-hand side and so makes the interpretation that 'longer distance means longer time', or else, 'higher speed means shorter time'.

Allied to the equation in t, d and v are others taken from physics which Piaget uses in connection with psychological time. Thus, the relation

$$\text{time} = \text{work done} \div \text{power}$$

finds an analogue in the observation that the estimated duration of an activity increases with the work accomplished, but decreases when the directed activity increases.

Piaget's conclusions have frequently met with opposition. Paul Fraisse, in particular, disagrees regarding time perception. He asserts [261], 'In our opinion the young child has intuitions not only of speed and distance but also of duration', and comments:

> 'One might almost say that he [Piaget] sought situations in which the relationship *time = distance/speed* was apparent. He affirms several times that the *notion* of time only takes shape when there is first articulate intuition and then the relating of speed and duration. We, on the other hand, tend more toward the idea that the representation of a duration, being already of an abstract nature and constituting a ground on which several changes are located, appears before the point at which the child becomes capable of relating order and duration by also taking into account the logical connections between time, distance and speed. We also do not believe that life, which gives the child occasion to meet the resistance of time, presents it to him as a relationship between the work accomplished and the speed of the action.'

Fraisse believes that there is an intermediate stage in the child's development in which intuition is gradually transformed into increasingly abstract representations, and the representation of duration becomes more and more independent of what happens in it. Despite such differences of opinion, however, it seems evident that at the later childhood and adult stages the relation $t = d/v$ is an important one in the judgement of duration.

We turn now to the *accuracy* of man's judgement of duration, under normal, favourable conditions. Experimenters make use of two main kinds of interval. These are:

(i) The 'empty' interval, which is bounded by two stimuli such as light flashes or audible clicks.

(ii) The continuous or 'filled' interval during which a light or sound (or possibly a touch) is maintained continuously.

Behaviour with regard to these types of interval is not always the same, and even the judgement in one type will depend on the stimulus used.

When the interval is too short it is not recognized by the subject as having duration at all. He is aware only of a single instantaneous flash or sound. In case (i), the bounding stimuli appear separate when the duration reaches or exceeds a critical value called the *threshold of the perception of succession*. In case (ii), time is not perceived until the *threshold of duration* is reached.

For a continuous light stimulus, the threshold of duration is of the order of 0·1 second, and depends on intensity. But for a sound stimulus it is only about 0·01 second (because the mechanical processes of the ear are quicker than the photochemical ones of the eye) (Durup and Fessard, quoted by Fraisse). Tactile stimuli give rise to figures close to those for sound.

The threshold of the perception of succession depends on the nature of the stimuli, and systematic effects (which are often quite consistent in each individual but vary from person to person) are encountered if the first stimulus is of a different kind from the second. (A similar effect produces errors of the order of 0·1 second in judging the position of a moving pointer on a dial at the moment of occurrence of a sound. In astronomy, a 'personal equation' has to be applied to the visual observations of each observer to allow for his own systematic error in tapping at the instant a star crosses a hairline in the telescope.) When two stimuli of the same kind are used, the threshold figures are about the same as those for duration, i.e. about 0·1 second for light and 0·01 second for sound and touch.

Much shorter intervals can be detected in a long succession of stimuli, which the subject then observes as a continuous stimulus varying in intensity. George Miller and Walter Taylor at the Psycho-Acoustical Laboratory at Harvard University performed experiments in 1948 with 'white noise', which is a mixed sound consisting of all audible frequencies [283], and found that the ear could detect regular interruptions at the rate of 1000 to 2000 per second. But, as we have stated, in this experiment the subject is not aware of an actual interruption and perceives only a variation in intensity in the signal.

To sum up, therefore, we may say that for the true recognition of an interval, man's threshold occurs at about 0·01 second and that this applies when the senses of hearing and touch are called into play.

Let us consider now the actual estimation of durations. In this connection, those roughly of the order of one second prove to be of particular interest. If audible stimuli are separated by an interval exceeding the threshold but less than about 0·6 second, one is less conscious of the duration itself than of the limits. According to Fraisse, 'we do not spontaneously perceive a gap. We perceive two more or less closely linked stimuli. The interval is not perceived in itself, although it is discernible if we fix our attention on it.' Between 0·6 second and 1 second the interval and its limits appear as a unit. For longer gaps the interval is predominant and the stimuli appear quite separate. 'When the gap reaches 1·8 to 2 seconds, the two stimuli cease to belong to the same present; we no longer perceive one interval of duration but only the distance between a past event and a present event.'

The three types of interval in which, respectively, the subject is primarily conscious of: (i) The limits alone. (ii) The limits together with the gap. (iii) The gap itself, are conveniently classified as 'short', 'indifferent' and 'long'. The classification is important in that short intervals are found to be systematically overestimated and long ones underestimated, the greatest accuracy occurring in the intermediate 'indifference zone'. Usually a subject is asked to estimate an interval by comparing it with another interval (used as a standard), or by reproducing the given one by tapping on a key. Thus, in the reproduction method the subject will normally tap out a longer interval than the given one, if the latter is 'short', and will similarly tap out too short an interval if the given one is 'long'.

Many determinations of the indifference zone, the zone of greatest accuracy of estimation, have been made and it is now usually quoted as about 0·6 to 0·8 second. Various factors can alter the determination.

Practice, even that occurring during a long experimental session, has a pronounced effect so that by the end of the session the subjects are best able to estimate intervals closer to the average of those used in the tests.

The existence of an indifference interval appears to have an absolute significance, and this could perhaps be one of the most important facts in respect of man's conscious time-keeping. Woodrow [305] refers to the possibility that the terms short and long (despite relative connotations) may be meaningful in an absolute, qualitative sense. Attempts have been made to determine the lengths of time to which they apply. The point at which the subject's judgement changes from short to long is the *absolute judgement indifference interval* and appears to be somewhere between 0·6 and 0·7 second, though it may be substantially affected by presenting the subject with shorter or longer intervals beforehand. Benussi (quoted by Woodrow) found, for example, that the change from short to long appeared to occur at 0·23 second when subjects were presented with a sequence of intervals of increasing length, ranging from 0·09 second to 2·7 seconds, whereas it occurred at 1·17 seconds when the same intervals were presented in decreasing order. For random ordering, the changeover value was between 0·58 and 0·72 second. The subject seems, therefore, to make use of recently experienced standards, *but there is nonetheless an inherent stability*. In Woodrow's view:

> 'There is no conclusive evidence, then, to rule out the possibility that the "standard" for the absolute judgement of long and short may be a relatively stable one, corresponding to an absolute judgement indifference level.'

It has variously been pointed out that the zone around 0·7 second is related to specific physiological processes. For example, Wundt has noted, '[¾ second] is almost the same as that taken for the swing of our leg when we are walking quickly. It seems not unlikely that this psychic constant for the mean duration of reproduction and the most accurate estimation of intervals has developed under the influence of body movements', and Guyau, 'Even today we still adapt the speed of our representations to the rhythm of walking.' (See [261].) Fraisse lists several examples of the occurrence of this interval, in reactions to various stimuli, in the identification of groups of words and numbers, and in the heartbeat. He suggests that these various durations are not regulated by one another, but that all the phenomena correspond to an optimum rhythm for successive associations in the nervous system.

Now let us turn to the consideration of somewhat less conscious time regulation in living beings, and especially in connection with the measurement of longer time-intervals than we have discussed so far. Many areas of study have shown the ability of all manner of organisms to record accurately the passing of several hours or days. Bees use inner clocks in their navigation when searching for food. In one notable experiment, Max Renner of Munich University demonstrated this convincingly. In Long Island, New York, bees were trained to fly to a feeding station in the north-west at 1 p.m. Then the bees (in their hive) were transported by air to California. Next day, the bees set off from the hive at 10 a.m. local time (1 p.m. New York time), and flew in a south-westerly direction for their food. In this direction, the sun was in the same relative position to their flight path as it had been on their accustomed journeys in New York. Therefore, their time-keeping was internal and independent of the sun's position which, however, was used as a compass. (Source: Frank A. Brown [250].)

Bees are also capable of taking into account the changing position of the sun during the day. If, for example, they are placed in a light-proof box while out feeding and are later released, they will return accurately to the hive. The inner clock guides them as to the new position of the sun.

A similar, remarkable, ability to navigate is shown by migrating birds. These use inner clocks in conjunction with the sun as a compass by day and the stars as a compass by night. Hamner [264] describes an instance in which migrating birds were caught and placed in a planetarium. They then showed a strong indication of wanting to fly in a particular direction relative to the stellar pattern in the ceiling. As the pattern was rotated, so the preferred direction changed. The birds, however, made allowance over a period of time for the slow rotation of the stellar pattern that would ordinarily be observed in the course of a night.

A second instance of biological time-keeping is found in the phenomenon of *photoperiodism,* in which the behaviour of plants and animals is seen to depend on the relative length of day to night. Certain plants will flower only when the day is short, others only when the day is long. Short-day plants are observed in some cases not to flower in continuous light, but will commence flowering after the light is interrupted by a single long dark period (signalling a short day). The duration of the dark period is critical to within a few minutes, and its effect may be nullified by a brief spell of lightness. (See Hamner [265].)

A major part of the research into biological time-keeping has been concerned with the so-called 'circadian rhythms' (*circa* = about, *diem* = daily). Plants, such as bean seedlings, start to raise their leaves at about daybreak and to lower them again at dusk. This rhythmic behaviour is found to continue even when the plants are subjected to uniform conditions, with constant temperature and lighting. Furthermore, the process may be 'rephased' by means of a few appropriate periods of alternate light and dark. Thus, the leaves may be induced to rise at night and droop by day. The new rhythm then persists if the lighting is once again kept constant.

The natural cycle of the rhythm is not exactly twenty-four hours, and it may be shortened or lengthened by giving say ten hours' light followed by ten hours' darkness, or fourteen hours' light followed by fourteen hours' darkness, resulting in a new cycle of twenty hours or twenty-eight hours' in the respective cases. However, if uniform lighting conditions are restored the plant reverts to its original rhythm. Surprisingly, the natural cycle is very little affected by temperature changes, unless these are extreme. Freezing temperatures may stop the clock, so that rephasing of the cycle takes place. (Much of the study in this field has been made by Erwin Bünning; the reader is referred to his book, *The Physiological Clock* [252].)

The circadian rhythm is evident in all manner of organisms from fungi to fruit flies, and from carrot slices to crabs. The fiddler crab changes colour through the action of pigment in the skin, becoming dark at daybreak and pale at sunset. The accuracy of the twenty-four-hour cycle is very high even in conditions of uniform darkness and it is affected very little by temperature change, indicating that the cycle is not related to metabolism. If the crab is placed in ice water for a few hours, however, the clock is found to be rephased by approximately this amount.

Man's own circadian rhythm exhibits its own features of persistence. While the daily routine may be based artificially on a period other than twenty-four hours, the twenty-four-hour cycle in, for example, bodily secretions remains unchanged. Hamner [264] describes experiments at Spitzbergen, where the sun shines continuously in summer, in which the clocks, watches and the daily life of the colony were all adapted to a twenty-one-hour 'day', and on another occasion to a twenty-seven-hour 'day'. The subjects soon became superficially accustomed to the new timings, but their secretions continued to work at the original pace. Conflict between the two time-rates appeared to cause the subjects some stress.

Other circadian cycles in man occur in certain blood counts, the heart rate and the blood pressure, etc. Temperature effects are not great, because the internal body temperature changes little with ambient temperature. But when the body temperature increases, as in a fever, human time-keeping does seem to go 'faster', as some tests have shown.

Biologists still dispute the question of whether the biological clock is purely internal or is controlled by external influences. There are certainly hereditary influences. Professor Frank A. Brown [249] believes that an inflow of external factors governs the timing, and that various organisms are surprisingly sensitive to geophysical variations. According to Professor J. L. Cloudsley-Thomson [253], the earth's rotation is not a factor, however, because the circadian rhythm is maintained in plants and animals even at the South Pole. Most biologists apparently believe in the genuine 'inner clock'; some believe that each cell is, or contains a biological clock, and that the synchronization is performed by the central nervous system. The question is of importance. In Cloudsley-Thomson's view, 'if biological clocks do depend for their accuracy upon the receipt of geophysical information, the removal of this might have dire effects on the space traveller.'

We return now to the question posed at the beginning of this section, concerning the relation between the living clock and the standard clock of physics. It is clear that unless geophysical factors play a much greater part than almost all biologists think, man can survive without earth's time-keeping. He thus keeps pace (in his ageing as well as in short-term temporal processes) with bodily clock mechanisms which are subject to the usual laws of chemistry and physics. These laws are the same, according to special relativity theory, in all inertial reference frames. The laws expressed mathematically contain a time-variable, which is the standard time in the reference frame of interest. Thus, there is no other possibility than that human ageing is related to standard time, and that objections like that of L. O. Pilgeram (p. 20) are unfounded so far as inertial motion is concerned. In accelerated motion, it would seem likely that human clocks would behave much as any other clock, provided that the acceleration were not sufficient to cause bodily damage. In this respect man is, unfortunately, rather vulnerable.

3. Accelerated clocks

So far, in regard to relative ageing and time-keeping we have avoided any detailed consideration of accelerated clocks by restricting the discussion to space journeys of a particular type, in which the periods of acceleration are comparatively brief. In the calculation in §1 we simply made an appeal to the observed fact that modest accelerations do not radically affect the rates of clocks of suitable construction.

Einstein did not make this restriction in his 1905 paper, but dealt first with the case where a clock (C') moves along any polygonal path from a point A to another point B, both specified in an inertial reference system S. The possibility was allowed that B and A are coincident, so that C' then moves in a closed polygonal path. It was assumed that the speed v of C' is constant, and therefore along each straight stretch the clock would run slow in S by the factor $\sqrt{(1-v^2)}$, where as usual we have put $c = 1$. If T is the S time for the motion from A to B, the time recorded by C' for the trip is

$$T\sqrt{(1-v^2)} = T - \{1 - \sqrt{(1-v^2)}\}\,T.$$

Thus, if v is much smaller than the light speed 1, C' will appear in S to have lost by the amount

$$\{1 - \sqrt{(1 - v^2)}\}T = \tfrac{1}{2}v^2 T,$$

approximately.

Next, Einstein made a generalization to any closed path described at constant speed v by introducing an explicit assumption:

> 'If we assume that the result proved for a polygonal line is also valid for a continuously curved line, we arrive at this result: If one of two synchronized clocks at A is moved in a closed curve with constant velocity until it returns to A, the journey lasting t seconds, then by the clock which has remained at rest the travelled clock on its arrival at A will be $\tfrac{1}{2}tv^2/c^2$ seconds slow.'

In effect, Einstein's assumption was that the rate of a clock depends only on its *velocity* and not on its *acceleration* (although strictly speaking he was dealing only with the case where the speed v is constant). But this is by no means a trivial assumption, and we can devise time-keeping mechanisms which would record at very different rates over a closed polygonal path and a continuously curved path described at the same speed.

Suppose that we purchase an electric chronometer and modify it for use in a (non-relativistic) experiment to be carried out on earth. First we construct a switching device which stops the chronometer whenever a certain electrical contact is broken. The wiring is arranged so that this contact is between a steel ball, which rolls inside a plastic cup, and a small metal terminal at the bottom of the cup. The chronometer and cup are then fixed on a trolley which can be moved freely on a smooth floor. If the trolley is at rest or in uniform motion (at speeds where relativistic effects are negligible) the chronometer goes normally, at the same rate as other clocks on earth. But whenever the trolley is accelerated the ball rolls away from the terminal, and contact in the switching circuit is broken. Thus, in general, time is recorded normally during unaccelerated motion and not at all during acceleration.

If the trolley is moved over a polygonal path at constant speed there is a pause in time-recording at each corner, but otherwise the chronometer registers correct earth time. However, in a continuously accelerated path the ball may well be displaced the whole time and nothing whatsoever is recorded. This is an example of an 'acceleration-sensitive' time-keeper, and although it depends on gravity for its working it is possible to invent other similar devices which do not, and which are acceleration sensitive when used in space.

From this rather artificial example it is clear that acceleration, as distinct from speed, may well affect the rate of any particular clock, but that one cannot decide this without a knowledge of the mechanism. If the rate is governed, say, by a balance-wheel then the problem is a rather complicated dynamical one. But from experience, as pointed out in §1, there is evidence that such clocks are not much affected by accelerations of magnitudes up to a few g, in which case Einstein's assumption is reasonable.

The statement that the instantaneous rate of a (suitable) clock depends only on its instantaneous speed is known as the *clock hypothesis,* and is sometimes used in the definition of an 'ideal' clock. Mathematically, the hypothesis may be stated as follows. If $v(t)$ is the speed of a clock C' at time t in the inertial reference system S, and if $\Delta T'$ is the time recorded by it in the brief interval $(t, t+\Delta t)$ (during which v is approximately constant) then

$$\Delta T' = \sqrt{(1 - v^2)}\Delta t. \tag{3.1}$$

For the arbitrary motion of C' along any path, the total time it records is evaluated by dividing the path into elements and applying (3.1) to each element. A limiting process then leads to the integral formula for the total recorded time T',

$$T' = \int \sqrt{(1 - v^2)} \, dt \tag{3.2}$$

where the limits of integration are the initial and final values of t. This integral is known as the *proper time* along the relevant part of the world-line of C. An *ideal* clock is one which records proper time along its world-line.

There is a considerable advantage in adopting the Einstein–Langevin clock (Chapter 2, §4) as the time-standard because it is comparatively simple to discuss its properties during acceleration, provided that the clock is not too large. If the clock is large we have to be precise about the mode of acceleration, and complications can arise. Suppose, for example, that it is travelling along Ox in S (with its mirrors perpendicular to Ox) and that it has a certain constant speed v until forces on the bar supporting the mirrors cause acceleration. If only a single force is applied at one end A of the bar, then the other end B will not change speed immediately because this would imply that a 'cause' at A had resulted in an immediate 'effect' at B, and the most rigid substance conceivable can transmit causal influences only at the speed of light (Fig. 18). What happens is that

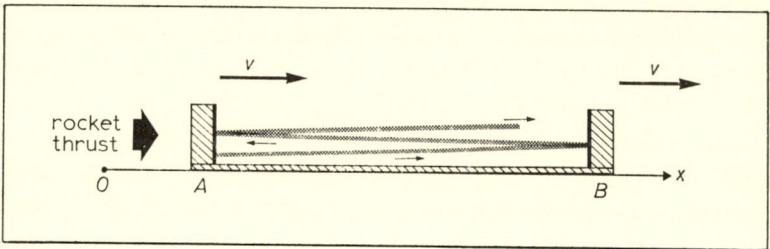

Fig. 18

a compression wave along the bar is set up by the applied force, and its effect on the length of the bar must be taken into account.

Such complications are not eliminated by assuming that simultaneous forces are applied at both ends (or at all points of the clock structure) since simultaneity is relative. If forces are applied simultaneously in S we get one type of behaviour; if they are applied simultaneously in the rest frame of the clock we get another, and so on.

When a thrust is applied only at the trailing end A (as might conceivably apply to a clock in a space-ship), the time of travel of a light ray which has just left A and is heading for B will be unaffected by the thrust. But a ray which has left B and is heading for A will have slightly less far to travel than if A had not been accelerated. This illustrates the type of difficulty that arises when too large a clock is used; the time-keeping depends on the stage of motion of the ray. If the distance AB is short, any speed changes at either end have negligible effect during the time a light ray takes to go from A to B or from B to A. Therefore, the clock hypothesis is valid for a sufficiently short Einstein–Langevin clock.

Let us turn now to the calculation of the passage of time during a journey in an accelerating space-ship. The most acceptable way to achieve high speed in space-travel from the passengers' point of view is by means of sustained gentle acceleration. We shall consider the particular case of relativistic *uniform acceleration*, which is the analogue of constant acceleration in Newtonian mechanics. It is not defined (as one might at first expect) by a condition such as 'acceleration in S equals constant' for this would not be a very useful definition. For one thing, constant acceleration in one inertial frame does not imply constant acceleration in all others in relativity, as it does in Newtonian theory. (The proof of this statement follows from equation (A.1) in the Appendix.) Therefore, the condition would have a different meaning for different observers. Again, if it were possible to sustain constant acceleration in any inertial system indefinitely the speed would eventually exceed c, which again indicates that the condition is unlikely to have much practical significance in relativistic mechanics. (In relativistic mechanics, the inertia of a body increases with its speed, the mass being given by the expression $m_0/\sqrt{(1-v^2)}$, where the constant m_0 is the *rest-mass*. The mass increases indefinitely as v approaches 1, and so no force could cause constant acceleration in a body indefinitely.)

The most sensible definition of relativistic uniform acceleration emerges from the following dynamical considerations. Imagine the space-ship to run its motors at a fixed setting of the controls, so that propellant is ejected at a constant rate and constant velocity relative to the ship. If, during the period of acceleration the mass of propellant used is negligible, compared with the total mass of the ship, its occupants and the remaining fuel, etc., then the motion of the ship should satisfy any reasonable criterion of uniform acceleration. It would certainly be one of constant acceleration according to Newtonian mechanics.

At any instant there is a particular inertial frame (whose speed in S is the same as that of the space-ship) in which the ship is momentarily at rest although it is accelerating. We call this a *co-moving* frame. A

93

different co-moving frame exists for each instant, and our definition of uniform acceleration is that the observed acceleration is the same in every co-moving frame. This follows because the space-ship, its controls, and the rate of ejection of propellant appear exactly the same in each of them.

What does the motion look like in the 'fixed' inertial frame S of earth? It is shown in the Appendix that if the acceleration in the co-moving system is a, then the acceleration in S is

$$\frac{dv}{dt} = a\beta^3, \tag{3.3}$$

where v is the speed in S and $\beta = \sqrt{(1-v^2)}$. Suppose that the speed is zero at time $t = 0$. By integration we get

$$\frac{v}{\sqrt{(1-v^2)}} = at,$$

which rearranges to give

$$v = \frac{dx}{dt} = \frac{at}{\sqrt{(1+a^2t^2)}}. \tag{3.4}$$

This shows that if the uniformly accelerated motion were sustained indefinitely, the speed v would progressively approach the light speed 1.

If we integrate again, and put $x = 0$ at $t = 0$, we find that

$$x = \frac{\sqrt{(1+a^2t^2)}-1}{a}. \tag{3.5}$$

The world-line for this motion is shown in Fig. 19. Also shown, for comparison, is the curve $x = \frac{1}{2}at^2$, which is the world-line according to Newtonian kinematics for the same constant acceleration and starting conditions. The broken straight line is an asymptote to the relativistic world-line and has the equation $x = t - 1/a$. It is a light-line because its slope is 1.

As an example, suppose that the space-ship sets off from earth to visit a distant planet, star or nebula. To make life comfortable for the passengers and yet attain as high a speed as possible the acceleration is kept constant for the first half of the outward journey, while for the

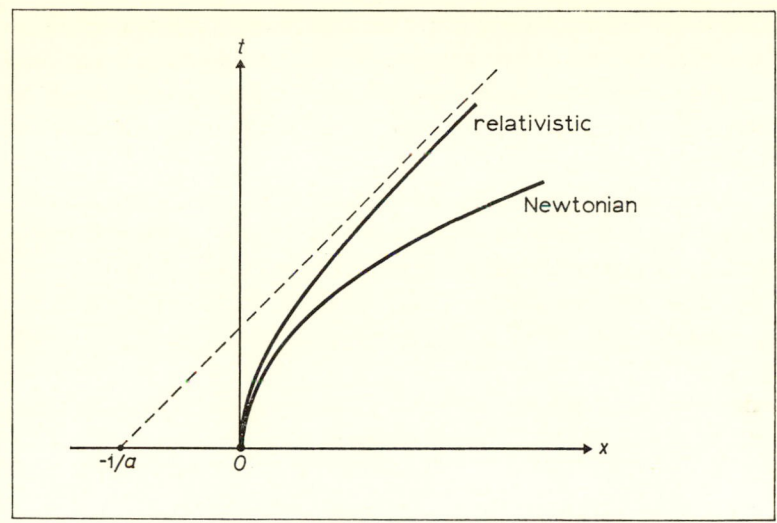

Fig. 19. *Uniform acceleration in Newtonian and relativistic kine-matics*

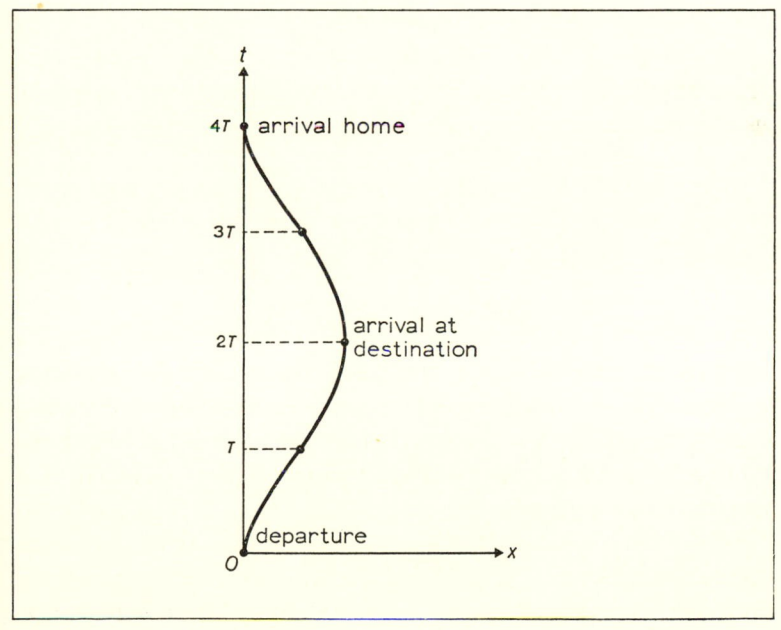

Fig. 20

second half a deceleration of the same constant magnitude is maintained (by reversing the rocket thrust). The return journey, made after a negligibly brief stay at the far destination, is the exact reverse of the outward one; the world-line of the whole voyage is thus as shown in Fig. 20.

Suppose that each period of acceleration or deceleration lasts for earth time T, the total earth duration of the voyage being $4T$. We shall calculate the duration according to passengers, using the clock hypothesis. By symmetry, this will be equal to $4T'$, where T' is the passengers' time for the first half of the outward journey. The clock hypothesis gives

$$T' = \int_0^T \sqrt{(1-v^2)}\, dt = \int_0^T \frac{dt}{\sqrt{(1+a^2t^2)}},$$

by (3.4). On carrying out the integration one finds that

$$T' = \frac{1}{a}\ln\{aT+\sqrt{(1+a^2T^2)}\}, \tag{3.6}$$

where ln denotes the natural logarithm.

The farthest distance reached (in earth measure) is twice that travelled in time T, and we get on putting $t = T$ in (3.5),

$$\text{farthest distance} = 2\frac{\sqrt{(1+a^2T^2)}-1}{a}. \tag{3.7}$$

Finally, the maximum speed attained is given by putting $t = T$ in (3.4). Thus,

$$\text{maximum speed} = \frac{aT}{\sqrt{(1+a^2T^2)}}. \tag{3.8}$$

Let us substitute some possible values. Suppose that for comfort, a is chosen to be equal to g, the gravitational acceleration at the earth's surface. Then, in our units in which length is measured in metres and time in $1/(3\times10^8)$ seconds (to make $c = 1$), we have numerically

$$a = \frac{9\cdot81}{(3\times10^8)^2} = 1\cdot09\times10^{-16}.$$

In Table 3 are shown figures calculated for various values of T. Columns 1 and 2 represent corresponding values of $4T$ and $4T'$, respectively, column 3 the farthest distance according to (3.7), and column 4 the maximum speed attained according to (3.8).

It is of interest to compare the fourth row in the Table with Sir George Thomson's calculation for a journey to the nearest star. His journey, at *constant speed* $\frac{1}{2}$, takes seventeen earth years as against

TABLE 3

Round-trip voyages with acceleration of constant magnitude 1g.

Earth time 4T	Space-ship time 4T'	Farthest distance reached	Max. speed / Light speed
4 days	4 days, less $\frac{1}{2}$ sec	73 million km (Mars?)	0·0028
3 months	3 months, less 91 min	35 light-hr	0·065
4 yr	3·5 yr	0·85 light-yr	0·72
11·7 yr	7·1 yr	4·2 light-yr (nearest star)	0·95
48·5 yr	12·5 yr	22·4 light-yr	0·9968
57·0 yr	13·1 yr	26·6 light-yr	0·9977
2350 yr	27·5 yr	1170 light-yr	0·999 998 6
4×10^6 yr	56·4 yr	2×10^6 light-yr (Andromeda)	$1 - \dfrac{1}{2 \times 10^{12}}$

rather less than twelve years at *constant acceleration* 1g. For the crew and passengers, the corresponding times are about fourteen and a half years and seven years, respectively, and during the accelerated trip the maximum speed attained is 95 per cent of the speed of light. Note also the enormous saving of time for travellers to Andromeda, who need only 56·4 years to survive four million earth years. However, those who take only the four-day excursion to Mars save a mere half second. Some of the other particular values in the Table are included for use in the next section.

For comparison, data for the case $a = 2g$ are also given, in Table 4. Here it is worth noticing that one travels little farther in a given period of earth time at acceleration 2g than at acceleration 1g. This is because in either case most of the trip is made at approximately the same speed, very nearly equal to 1. But the travellers' time to Andromeda and back is halved when the greater acceleration is used. The feasibility of such intergalactic journeys will be dealt with in the next section.

TABLE 4

Round-trip voyages with acceleration of constant magnitude $2g$.

Earth-time $4T$	Space-ship time $4T'$	Farthest distance reached	Max. speed / Light speed
4 days	4 days, less 1·8 sec	146 million km	0·0056
3 months	3 months, less 6 hr	70 light-hr	0·13
4 yr	2·85 yr	1·25 light-yr	0·90
11·7 yr	4·8 yr	5·0 light-yr	0·987
48·5 yr	7·6 yr	23·3 light-yr	0·9992
2350 yr	15·1 yr	1170 light-yr	0·999 999 66
4×10^6 yr	29·5 yr	2×10^6 light-yr	$1 - \dfrac{1}{8 \times 10^{12}}$

4. On visiting people elsewhere*

In two fascinating articles [230; 231] in *Science,* in December, 1961, and July, 1962, Sebastian von Hoerner of the National Radio Astronomy Observatory at Green Bank, West Virginia, reviewed some of the practical and theoretical limitations which face those who wish to contact or visit other civilizations in the universe. The first article, entitled 'The Search for Signals from Other Civilizations', was concerned with estimates of the longevity of a civilization once the advanced technical stage is reached, of the distance to the nearest stars whose planetary systems are likely to support such civilizations at the present time, and the possibilities of radio communication with them. The second article, 'The General Limits of Space Travel', dealt with the propulsion problem of covering the necessary distances, especially if use is to be made of time-dilatation.

Von Hoerner takes as starting point the assumption that 'anything seemingly unique and peculiar to us is actually one out of many and is probably average'. This sort of principle of 'typicality' has to be invoked constantly in science if any progress at all is to be made. In the absence of other information we naturally assume that certain

*Most of the material in this section is due to Sebastian von Hoerner, and was published in the two articles cited. I gratefully acknowledge his permission and that of the journal *Science* for its reproduction. (Copyrights 1961 and 1962 by the American Association for the Advancement of Science.) Some additional calculations are my own.

broad features of our environment and the development of our civilization are typical of any elsewhere, though we do not of course assume that other planetary systems and the life which they support must be like our own in matters of detail. (In a more modest way we rely on a similar principle constantly in daily life, whenever we visit new places or encounter new situations.)

It is thus reasonable to suppose that:

(i) Life and intelligence develop elsewhere by the same rules of natural selection given the proper surroundings and requisite time.

(ii) The average civilization will arrive at our present level of science, technology, and the desire for interstellar communication after about the same time as ourselves.

It is to be borne in mind that any particular civilization might well not last indefinitely; von Hoerner argues:

'Science and technology have been brought forward (not entirely, but to a high degree) by the fight for supremacy and by the desire for an easy life. Both of these driving forces tend to destroy if they are not controlled in time: the first one leads to total destruction and the second one leads to biological or mental degeneration.'

He thus assumes that similar states of mind to our own will have developed at many places but that each will have only a finite lease of life.

Let us calculate the probable average distance between neighbouring technical civilizations. For the few million or so nearest stars (just the 'local' part of our galaxy), we introduce the following average quantities:

T_0 = time needed to develop a technical civilization from the birth of a star.

T = age of oldest stars.

v_0 = fraction of all stars which possess planets capable of supporting life; these we call 'favourable' stars.

v = fraction of all stars which at present have a technical civilization.

l = longevity (i.e. life duration) of the civilization after the technical stage is reached, until destruction or degeneration occur.

Suppose that star formation is a fairly constant feature. Then the ages of existing stars will be evenly distributed up to the maximum value T. The proportion of stars older than T_0 is $(T-T_0)/T$, and if there are no longevity limitations we have $v = v_0(T-T_0)/T$. On the other hand, if l is less than $T-T_0$, then a proportion l/T of the favourable stars will possess technical civilizations at the present time, and $v = v_0 l/T$. Thus,

$$v = v_0 l/T, \qquad \text{if} \quad l \leqslant T-T_0, \qquad (3.9)$$
$$v = v_0(T-T_0)/T \quad \text{if} \quad l \geqslant T-T_0.$$

Von Hoerner considers five main cases of longevity limitation:

(1) Complete destruction of all life.
(2) Destruction of higher life forms only.
(3) Physical or mental degeneration and decay.
(4) Loss of interest in science and technology.
(5) No limitation at all.

Let $p_1, p_2, \ldots, p_5$ be the probabilities that any civilization suffers these respective fates, and let $l_1, l_2, \ldots, l_5$ be the respective longevities $(l_5 = T-T_0)$. The average longevity is

$$l = p_1 l_1 + p_2 l_2 + \ldots + p_5 l_5, \qquad (3.10)$$

and if each favourable star supports only one civilization in its lifetime, then

$$v = v_0 l/T, \qquad (3.11)$$

with this value for l.

If the longevities l_2 and l_3 are short enough, it is worth taking into account the possibility that a second civilization develops on the same planet in cases (2) or (3). This should occur in a fraction $p_2 + p_3$ of all favourable stars, and since any one of the above five fates might occur to the second civilization, a fraction $p_2 + p_3$ of stars which support a second civilization will also support a third one, and so on. Assuming that the recurrence time is negligibly short, (3.11) will be replaced by:

$$\nu = \nu_0 lQ/T, \tag{3.12}$$

where Q is a *recurrence factor* given by

$$Q = 1 + (p_2 + p_3) + (p_2 + p_3)^2 + \ldots = \frac{1}{1 - (p_2 + p_3)}. \tag{3.13}$$

The choice of numerical values to be assumed for the various longevities and probabilities in cases (1) to (5) is bound to be highly conjectural. Primarily these are important in the evaluation of l, and uncertainty in l is likely to have its worst effect in the determination of ν, which in turn is needed in the distance calculation. Fortunately, the distance estimate does not depend too critically on ν. Let D denote the mean distance between neighbouring technical civilizations, and let D_0 denote the mean distance between neighbouring stars of all kinds. If we consider two regions of space, one containing a given number n of technical civilizations and the other a similarly shaped, but smaller, region containing the same number n of stars of all kinds, then the volumes of the two regions will be in the ratio $1 : \nu$. The linear dimensions of the regions, however, will be in the ratio $D : D_0$, and by comparing volumes we get $(D/D_0)^3 = 1/\nu$, i.e.

$$D = D_0/\nu^{1/3}. \tag{3.14}$$

This equation, together with (3.12), shows that D depends on l only through the cube root of l. Thus, an error in l by as much as a factor 8 leads only to an error in D by a factor 2.

Rather pessimistically, one hopes, von Hoerner puts the probability that a civilization has unlimited longevity (except as limited by the life of the parent star) as zero, i.e. $p_5 = 0$. (The possibility that a civilization has spread by colonization of nearby stars is apparently excluded.) His other adopted values for the longevities l_i and probabilities p_i ($i = 1, 2, \ldots, 5$) are as shown in Table 5. The reader may well care to contemplate alternative values; note in particular the very high assumed probability (0·6) of extinction by the destruction of higher life after an average longevity of only 30 years.

For the moment, however, we proceed with the calculations on the basis of von Hoerner's figures; by adding the values in column 5 we get, approximately, $l = 6500$ years. From columns 4, $Q = 4$.

What value should one adopt for T? This is known more reliably than the quantities considered so far. The ages of the oldest stars in

TABLE 5

Von Hoerner's values for longevities and probabilities.

Case	Estimated range for l_i (years)	Adopted value l_i (years)	Adopted value p_i	$p_i l_i$ (years)
(1) Complete destruction	0–200	100	0·05	5
(2) Destruction of higher life	0–50	30	0·60	18
(3) Degeneration	$10^4 - 10^5$	3×10^4	0·15	4500
(4) Loss of interest	$10^3 - 10^5$	10^4	0·20	2000
(5) No limitation	At least $T - T_0$	$T - T_0$	0·00	0

our part of the galaxy can be estimated from theoretical stellar models and the observed luminosities, radii and masses of numerous actual stars. Bondi [247] gives the age as between 3×10^9 and 8×10^9 years measured from the beginning of the 'main sequence' stage, where the bulk of the energy that the star loses in radiation is supplied by thermonuclear conversion of hydrogen into helium. Before the main sequence stage there first occurs the initial condensation of parent matter, but the time-scale of this process – where many stars are formed at once – is difficult to estimate because the density of the matter is very low. The birth of a star in most evolutionary theories is taken as the beginning of the next stage when the radius has fallen to a definite value. From then, until the main sequence is reached, the energy radiated is supplied by the release of the star's own gravitational energy during the further contraction, but this lasts for only a comparatively short time in the star's life, say between 10 million and 100 million years in the case of a star like the sun. Quite possibly Bondi's lower figure is rather too low; some estimates put the age of the earth as $4·2 \times 10^9 - 4·8 \times 10^9$ years (Allen [243]), in which case the sun would be at least this old.

There are numerous well-known stellar models from which an estimate of the total expected lifetime of a star can be made. For example, Menzel, Bhatnager and Sen [279] give details of a calculation for the sun using a homogeneous model due to Schwarzschild *et al.*, and find that the expected lifetime is about 10^{11} years. This gives a rough upper limit to the age. Apparently stars with masses close to that of the sun have not been observed at the terminal stage (H. Spencer Jones [301]). Spencer Jones puts the value of T at about

$$T = 10^{10} \text{ years},$$

which is close to Bondi's higher figure and to most other estimates. This is the value adopted by von Hoerner, who takes also

$$D_0 = 2 \cdot 3 \text{ parsecs*} = 7 \cdot 50 \text{ light-years}$$

as the average distance from the sun of the ten nearest stars.

His suggested value for ν_0, the fraction of all stars which are 'favourable' stars, is 0·06, but a minor arithmetical error causes subsequent calculations to correspond to $\nu_0 = 0\cdot1$. To follow through von Hoerner's study we shall adopt the latter value, which does not seem unduly optimistic according to present-day beliefs, though it would have been thought so not too many years ago. For instance, in 1930, Sir James Jeans [272] wrote:

> 'It is so unusual an accident for suns to throw off planets as our own has done, that only about one star in 100,000 has a planet revolving around it in the small [temperate] zone in which life is possible.'

Temperature is of course a critical factor for living organic material, and this severely limits the possible orbits of life-supporting planets. In 1961, Jagjit Singh [297] estimated:

> 'Barely 10 per cent of the stars in our Milky Way are born single and not every such star has a planetary system, and further that only 10 per cent of its planets may acquire the right blend of mass, axial rotation, distance [from the central star] and other attributes likely to favour the emergence of life and intelligence.'

*1 parsec = 3·26 light-years.

More promising are the assurances of Sir Bernard Lovell [275] (BBC Reith Lectures, 1958) that:

'Modern cosmogony can accept a situation in which most of the stars in the Milky Way have planetary systems similar to our own',

and of Fred Hoyle [269] that:

'We may expect planetary systems to have developed around the majority of the stars . . . Nor do the compositions of the planets seem in the least to be a matter of chance. Rather do I think it would be somewhat surprising if anything very different had occurred in any of the other planetary systems . . . Living creatures must it seems be rather common in the universe.'

G. Cocconi, Professor of Physics at Cornell University, shared the belief that a high proportion of stars possess life-supporting planets when he wrote to Lovell in 1959 suggesting a search for signals from them by means of radio telescope [276].* Cocconi thought that there is a good chance that among the 100 stars closest to the sun, some have planets with life well advanced in evolution. This fits in well with a proposed value for v_0 of at least 0·1.

From the above values of Q, l, v_0 and T we get by (3.12) an estimate for the proportion of all local stars which at this time possess advanced technical civilizations,

$$v = 2\cdot6 \times 10^{-7},$$

or 1 star in 3 or 4 million. By (3.14), with $D_0 = 7\cdot50$ light-years, we then obtain

$$D = 1170 \text{ light-years}. \tag{3·15}$$

(The alternative value $v_0 = 0\cdot06$ gives $v = 1\cdot56 \times 10^{-7}$, $D = 1390$ light-years.) A further analysis using these figures reveals that the most probable 'technical age' of the first civilization encountered is 12000 years (i.e. that it will be 12000 years beyond the stage of

---

*Such a search for signals was started on April 11, 1960, directed by Frank D. Drake of the National Radio Astronomy Observatory, USA, and given the code-name Project Ozma.

acquiring advanced radio techniques) and will therefore be one of those with longer than average survival time.

We have noted that D does not depend critically on v, although it does depend critically on D_0 (which however is known fairly accurately). The main source of error in v is likely to be due to a bad estimate of l, which involves the very subjective estimates of l_i and p_i. With so little direct evidence to hand, it would be futile to attempt to 'improve' on von Hoerner's values for the latter, but it is of interest to consider how much v and D change if one replaces pessimism with considerable optimism.

Suppose that a civilization has a 50:50 chance of surviving indefinitely. Then $p_5 = \frac{1}{2}$. We shall need to estimate T_0, which enters into the calculation of l, now that p_5 is non-zero. In the case of the sun, it has taken approximately $\frac{1}{2}T$ (i.e. $\frac{1}{2} \times 10^{10}$ years) for the technical stage to emerge on the planet Earth, and so we put $T_0 = \frac{1}{2} \times 10^{10}$ years.

The value of Q must now lie between 1 and 2, let us say for definiteness $Q = 1.5$. The exact values of $p_1, \ldots, p_4$ are elsewhere quite unimportant unless the corresponding longevities $l_1, \ldots, l_4$ in Table 5 are in error by several powers of 10. We now get

$$l = 2.5 \times 10^9 \text{ years,}$$

which is of course very much greater than von Hoerner's value. With $D_0 = 7.50$ light-years and $v_0 = 0.1$ as before, we find:

$$v = 0.0375,$$

or rather less than 4 per cent, as the proportion of all stars which have technical civilizations at the present. For the mean distance to the ten nearest of these, we have

$$D = 22.4 \text{ light-years.}$$

What is worth noting is that while this is very much smaller than (3.15), it is still more than five times the distance to our nearest stellar neighbour, Proxima Centauri. Although we can reach this star by making a journey of 4.2 light-years, it is probably necessary to go many times this distance to be received by highly intelligent hosts. If we make the trip successfully, however, our hosts are likely to be very wise indeed, and well worth meeting.

The theoretical (and not merely technological) limitations on high-speed long distance journeys are discussed in von Hoerner's second article. In fact, technological factors are the less important to our ability to engage in social life on a galactic scale, since our civilized neighbours will have probably overcome these factors, and we might therefore receive visits even if we cannot make visits ourselves. At present, our best knowledge of propulsion techniques for rocket ships relates to propulsion by the acceleration of exhaust materials by combustion. The limitations here are two-fold, in the energy content of the fuels used, and in the heat resistance of the combustion-chamber and the nozzle materials. The burning of hydrogen with oxygen, for example, gives about three times the energy per kilogram as TNT, but even by this method of propulsion it requires several kilograms of fuel to remove one kilogram of matter from the earth's gravity field. Figures for hydrogen-fluorine are comparable. Beyond the earth's gravity the method cannot be used to achieve really high rocket speeds, as the following arguments show.

The maximum temperature that nozzle materials can withstand is about 4000° C, and combustion gas at that temperature has an exhaust velocity of not more than 4 km/s. This limiting velocity would prohibit the use of fuels with extremely high energy content, if such were available.

A guide to the magnitudes of velocities attainable with various nozzle velocities and payloads can be determined by a non-relativistic calculation. Suppose that the rocket ship is beyond gravity and starts up from rest with initial mass (including fuel) M_i. Let the nozzle velocity s be constant, and let the final mass of the ship when all fuel is used be M_f. By considering the conservation of momentum during the expulsion of each element of fuel, one can obtain a well-known formula for the terminal rocket speed V in terms of the *mass-ratio* $M = M_i/M_f$,

$$V = s \ln M. \tag{3.16}$$

This relation shows that the final speed V of the rocket cannot be many times greater than s. When M_i and M_f are nearly equal, V is much smaller than s. But even if as much as 99·9 per cent of the initial mass were due to fuel (which is hardly likely), then $M = 1000$

and so $V = 6\cdot9s$. If all but 1 *part in a million* of the initial mass were fuel (the remainder has to include ship, occupants and other contents, and motor) we still get only the terminal speed $V = 13\cdot8s$, and so on. Therefore, extremely high-speed rocket ships using chemical fuels are not at all practicable, and the speeds feasible make relativistic calculations quite unnecessary.

Efficiency considerations confirm that a high nozzle velocity is

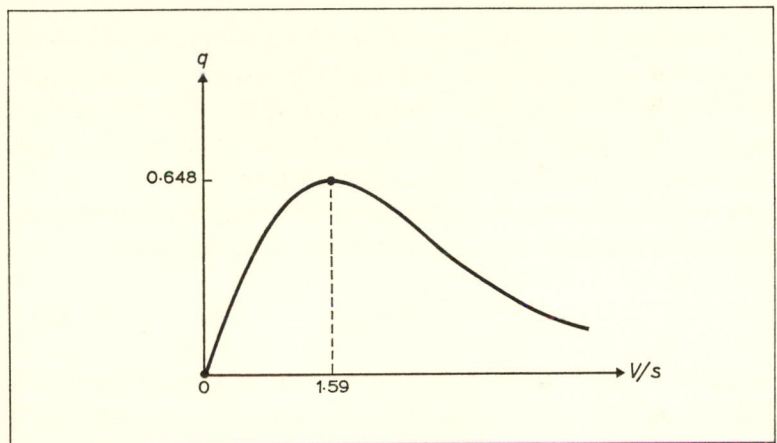

Fig. 21

needed if a high terminal speed is to be attained. The energy content of the burnt fuel is $\frac{1}{2}(M_i - M_f)s^2$ and the 'useful' energy (that in the final motion of the empty rocket) is $\frac{1}{2}M_f V^2$. The ratio, q, of the second of these energies to the first, is a simple measure of the efficiency of the rocket. By (3.16) we find:

$$q = \frac{V^2}{s^2(e^{V/s}-1)}.$$

A graph of q plotted against the ratio V/s is shown in Fig. 21. There is a maximum value $q = 0\cdot648$ when $V/s = 1\cdot59$, and q falls considerably when V/s is much greater than this. For example, the efficiency is only $0\cdot01$ for $V = 9s$. Thus, for $s = 4$ km/s, V cannot exceed 36 km/s without the efficiency dropping below 1 per cent.

Other types of propulsion are better. Uranium fission yields about 6 million times as much energy per kilogramme as the burning of hydrogen. The fusion of hydrogen into helium (as in the hydrogen bomb) gives another factor of 10. The use of ion thrust, in which charged particles are ejected by means of an electric field is another possibility, but is practicable only for low-thrust propulsion since s increases with only the square root of the field voltage and really enormous voltages are necessary to make s at all comparable with the speed of light.

The most conceivable energy obtainable for a given mass of fuel is obtained in the complete annihilation of matter with antimatter, resulting in radiation. This process yields about 140 times as much energy per kilogramme as hydrogen fusion, but seems a quite impracticable process for space-ship propulsion.

When relativistic considerations are important, at high speeds, the final velocity in one-way travel is given by the formula:

$$V = \frac{1 - M^{-2s}}{1 + M^{-2s}} \qquad (3.17)$$

in place of (3.16), where M is now the ratio of the initial and final *rest-masses*. For the best conceivable method of propulsion, by photons produced in an annihilation process, $s = 1$, in which case we get by rearranging (3.17):

$$M = \sqrt{\frac{1+V}{1-V}}. \qquad (3.18)$$

To see the sort of limitation that there might be on space journeys, we contemplate three journeys by photon rocket to stars in our part of the galaxy at different distances:

(i) 1170 light-years, von Hoerner's estimate of the average distance of the ten nearest technical civilizations in his first article (a somewhat lower figure was considered in the second article),

(ii) 22·4 light-years, our own 'optimistic' estimate of the same quantity, and

(iii) 4·2 light-years, the distance to the nearest star other than the sun.

To apply (3.18), let us simply consider the first half of each outward journey at constant acceleration $1g$, as described in the last section. The problem of saving fuel for the deceleration needed to complete the outward trip, and for the return (unless our hosts generously provide this) will be ignored for the moment. The relevant speeds V are given in Table 3, and if M is interpreted as the ratio of the initial total rest-mass of the ship to the total rest-mass at the half-way-out stage, we find by (3.18) the following values in the three cases (with c again put equal to 1):

(i) $V = 1 - 1 \cdot 4 \times 10^{-6}$, $M = 1200$.
(ii) $V = 0 \cdot 9968$, $M = 25$.
(iii) $V = 0 \cdot 95$, $M = 6 \cdot 2$.

Therefore, in case (i) all but 1 part in 1200 of the initial rest-mass must be fuel, while in (ii) the fuel accounts for all but 1 part in 25, and in (iii) it accounts for all but a part in $6 \cdot 2$. (Remember that this is for the best conceivable fuel.) And if the remainder of the outward journey is taken into account, then, even in case (iii), the rest-mass of the ship on arrival at Proxima Centauri can only be $1/(6 \cdot 2)^2 = 1/38$ times the initial value. Things are just beginning to look realistic. A modification of case (iii) allows longer trips by the inclusion of arbitrarily long periods of 'coasting'.

J. R. Pierce [188] has considered the feasibility of scooping up interstellar material to replenish the fuel supply. Most of this material is hydrogen distributed very irregularly in the form of clouds, the overall density in our own galaxy being of the order of 10^6 atoms per cubic metre (Allen [243]). Pierce calculates that if the ship were to use a scoop 10 000 square metres in area the amount of fuel collected would still be quite negligible from the point of view of high-speed travel. As an example, he finds that if this were the sole fuel used, and if the density of interstellar hydrogen were 1000 times as great as it is believed to be, then a space-ship of mass 15 700 kg (15·5 tons) would attain a speed of only 0·093 times that of light after travelling ten light-years.

Freeman J. Dyson, who worked on the Orion Project at San Diego in the late 1950s to design a space-ship propelled by nuclear explosions, considers hydrogen bomb propulsion feasible [256]. Large effective exhaust velocities are involved, some thousands of times as great as in chemical propulsion, giving rise to vehicle speeds of a few per cent of the speed of light. Assuming a long-term expansion of the U.S. Gross National Product at the rate of 4 per cent per annum, in order to meet the cost, Dyson predicts that 'about 200 years from now, barring a catastrophe, the first interstellar voyages will begin'.

In such early voyages, time-dilatation would not play a significant part. It is too soon to say whether man will find the means (or the desire) to explore the more distant reaches of our galaxy, except by exchanging radio signals. If he does engage in 'star-hopping', then the cumulative effect of time-dilatation will, of course, be a factor to consider. But here we should still think in terms of modest savings of time rather than the compression of centuries into seconds.

Chapter 4

THE DOUBTERS

'There are no doubt minds which have not this predisposition
to regard as substantial the things which are permanent;
but we shut them up in lunatic asylums.'

SIR ARTHUR EDDINGTON, *The Mathematical Theory of Relativity.*

1. The problem of acceleration

In this chapter we shall review a number of the arguments that have
been put forward 'for' and 'against' in the clock paradox controversy
in special relativity. Frequently, it has appeared that disputes have
continued simply because the answers given by one side have not
been confined to points raised by the other, and occasionally because
they have not been strictly relevant. Arzeliès [2] has criticized this in a
colourful way:

'When there is a discussion, the same arguments are always
advanced, and the same replies given . . . I suggest that the objections and replies might be labelled with the letters $A, B, \ldots$; if a
non-relativist opened box A, a relativist would press button
$B \ldots$'

On other occasions *ad hoc* rules have been introduced rather arbitrarily to deal with specific issues. These rules might take such forms
as 'A clock must satisfy such-and-such a condition' or 'To read a
clock you must be moving with it', and have thus provided opponents
with much ammunition.

Bones of contention may be roughly grouped under a number of
headings; these are the competence of special relativity to deal with
accelerated clocks and observers, the erroneous supposition of the
relativity of *all* motion, the synchronization of clocks and of simultaneity at a distance, the limitations on what constitutes a legitimate

clock, and so on. Naturally, there are often overlappings of these groupings. We shall start by considering some arguments involving acceleration.

An attempt to avoid the issue of acceleration completely has been made by some authors (e.g. Lord Halsbury [114]) by modifying the twin problem so that it becomes a 'problem of three brothers'. Brother A is at rest in an inertial system (on earth, say) and brother B travels directly away from him at high constant speed V. B's clock agrees with A's at the instant the two separate. Later, B passes brother C who is approaching earth at constant speed V (the three world-lines of the brothers are shown in Fig. 22), and B and C note that their two clocks agree as they pass each other. When C reaches earth, how does his clock reading compare with A's?

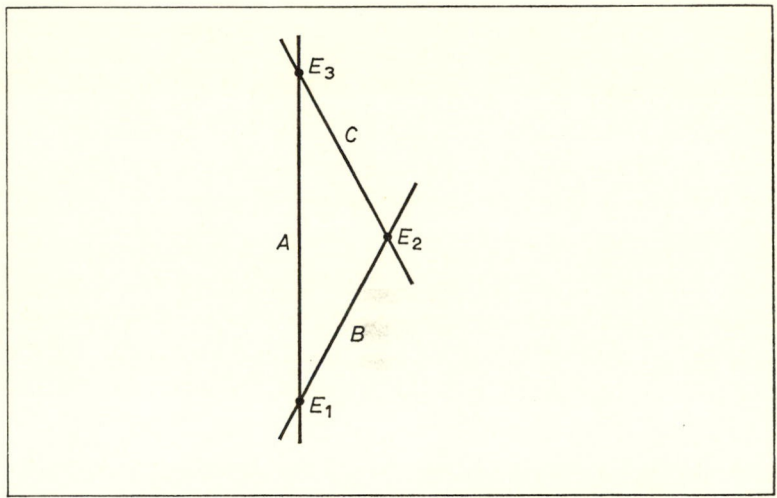

Fig. 22. *Lord Halsbury's problem of the three brothers. When B and A separate their own two clocks agree. When C passes B, their clocks agree. How does C's clock compare with A's when these last two meet?*

This problem can be treated quite rigorously by applying the Lorentz transformation equations to the various pairs of inertial systems. Suppose that A, B and C are at rest at the origins of parallel inertial frames denoted by S, S' and S'' respectively. Let the common

x direction be the direction of B's motion relative to A. If the event E_1 of initial separation occurs at time zero in both S and S', then the coordinates (x, t) and (x', t') of any event on the common x axes in these two systems are related by (2.9),

$$x' = \beta(x - Vt), \qquad (4\cdot1)$$
$$t' = \beta(t - Vx),$$

where $\beta = 1/\sqrt{(1 - V^2)}$, and as usual $c = 1$. Suppose that C passes B at time $t = T$. Then event E_2 is at $x = VT$, $t = T$ in S, and by (4.1) it is at $x' = 0$, $t' = \beta(T - V^2T) = T/\beta$ in S'. It follows, by assumption, that T/β is also C's clock reading at event E_2.

The Lorentz transformation between S'' and S differs in form from (4.1). First, V must be replaced by $-V$, and secondly the form (4.1) relates to the case where the origins of the two frames concerned coincide at the common zero of time, which is not the case for S'' and S. Instead, we note that event E_2 has coordinates (VT, T) in S and $(0, T/\beta)$ in S'', and the implied shifts of zero-points of space and time coordinates to comply with this give the required transformation:

$$x'' = \beta\{x - VT + V(t - T)\},$$
$$t'' - T/\beta = \beta\{t - T + V(x - VT)\}. \qquad (4.2)$$

Now, the meeting, E_3, of C and A occurs at the point $x = 0$ and the time $t = 2T$ in S. From the second of (4.2), the time of this event in S'' (and also C's clock reading) is

$$t'' = \frac{T}{\beta} + \beta(1 - V^2)T = 2T\sqrt{(1 - V^2)}.$$

This shows that the total time recorded by the clocks of B and C between events E_1 and E_3 is less than that recorded by the clock of A, by the factor $\sqrt{(1 - V^2)}$.

In terms of ageing, the combined ageing of the brothers B and C is less between these two events than the ageing of A, in agreement with the result in §1 of Chapter 3 for one travelling brother.

A common objection to this type of argument is that more than one S clock is used ($t = T$ is the reading of a local one at the event E_2) and that time-at-a-distance is therefore involved (as it is, legitimately, in the Einstein synchronization of S clocks). We have considered this point in Chapter 1, §5, and will meet it again in the next section.

We return now to some of the arguments concerning acceleration in the original problem of twins. Cullwick [45] maintains that 'even if the acceleration of an ideal* clock affects its rate *during the acceleration* [his italics], such a change of rate would not remain permanently after the acceleration has ceased, and so could not account for the alleged time difference'. He continues:

'The supporters of the claim [of asymmetrical ageing] are therefore left with Lorentz's original concept of an effect due to absolute motion through the aether. Actually they base their argument, usually, on an invalid application of Minkowski's four-dimensional exposition of the Lorentz transformation. We conclude, therefore, that both the paradox and its alleged solution are fallacious.'

In an appendix (in the first edition of his book) Cullwick examined in more detail the behaviour of moving clocks and deduced that while receding clocks run slow, approaching ones run fast. The argument, however, concerns only the observed reading of a distant moving clock by means of reflected light rays sent out by, and returned to, one observer. The variable time of travel of successive light rays gives rise to an effect identical to the Doppler effect, and in fact Cullwick's formulae are the relativistic Doppler formulae. But they are not interpreted by him as such.

Dingle feels that many of those holding what we have called the 'conventional' viewpoint are inconsistent and not generally in agreement with one another over the part played by acceleration. In one article [58] in 1956 he argued (referring to Sir George Thomson's example):

'To sum up, if accelerations are ignored, symmetry shows that the clocks cannot differ on reunion. If, however, accelerations

---

*Some authors, e.g. Rindler [290], use the term 'ideal' to signify a clock which records proper time, i.e. obeys the clock hypothesis. Others use it less restrictively in the sense in which we use 'standard'. Cullwick intends the latter meaning.

determine the issue, then either the $2\frac{1}{2}$ years [saved by the traveller] (if gained at all) is gained during them or they affect the later rate of the uniformly moving clock. The former's impossible; the accelerations may be far too brief. Hence the rate of a uniformly moving clock, for no known reason, depends on past accelerations, and all astronomical deductions from supposed Doppler effects are vitiated.'

The reference to Doppler effects here is, apparently, that if past unknown accelerations affect the time rates of stars, causing shifts of their recognized spectral lines, then the astronomical interpretation of an observed shift toward the red end of the spectrum as being due to a recession of the star concerned would be unwarranted. All previous (and unknown) accelerations would also have to be taken into account before any conclusion could be made. However, this argument bears no weight. The total effect of past accelerations is significant, in the present context, only in as much as they determine the velocity of the emitting star relative to the astronomer.

In another article, in the following year, Dingle referred again to the problem of acceleration [60]. Builder had previously emphasized [18] the dynamical asymmetry of the twins' motion, and Dingle was replying. (The notation M (= moving) has commonly been used to denote the space travelling twin or 'observer' and R (= rest) to denote his stay-at-home brother.)

'But in what sense is M accelerated rather than R? ... If, therefore, in spite of having to ignore the accelerations we reintroduce them for ulterior reasons, we must regard M as accelerated with respect to R, for there is nothing else in the problem to which to relate it. But if M is accelerated with respect to R, then R must be accelerated with respect to M; any other possibility is inconceivable. Hence the statement "M is the accelerated observer" is meaningless. It can acquire a meaning only if we decide to take into account the "something" that has "happened" to M; for example, if M is projected by an explosion or a gravitational field or something like that, then that, it is true, enables one to distinguish M from R, but it does not allow one to distinguish the *motion* of M from the *motion* of R.'

This is really the heart of the matter as far as the question of the symmetry of the motions is concerned. If there were no other bodies such as stars and interstellar matter in the universe (and at first sight it might appear that special relativity does not lean on the existence of these) it would be hard to make a case for the statement that M accelerates in a sense in which R does not. There would be nothing to distinguish any one type of motion from any other type, and only the relative motions of the two bodies would be meaningful. But one cannot make sensible statements about what the universe would be like if all the matter apart from the two observers were absent, and the centuries of observations which eventually led to the formulation of special relativity theory have all been made in our universe as it is, or then was. These observations have shown that a definite set of motions, the inertial motions, of which one is approximately the state of rest relative to the average motion of stars in our galaxy, are distinguished from all others in certain respects. Special relativity does not therefore *deny* the existence of the background matter of the universe. It merely regards whatever role that matter might have in determining the inertial motions as outside its province, and takes the observed existence of these preferred motions as a starting point.

On more than one occasion Dingle [62; 64] has put forward the following 'syllogism' in support of his viewpoint:

'(1) According to the postulate of relativity, if two bodies (for example, two identical clocks) separate and reunite, then there is no observable phenomenon that will show in an absolute sense that one rather than the other has moved.

(2) If, on reunion, one clock were retarded by a quantity dependent on their relative motion and the other were not, that phenomenon would show that the first had moved and the second had not.

(3) Hence, if the postulate of relativity is true, the clocks must be retarded equally or not at all; in either case, their readings will agree on reunion if they agreed on separation.'

The flaw in this argument is in (1), according to our comments above. There is no reason why the syllogism, if it applies at all, should not apply in suitable experiments with a pair of acceleration-sensitive clocks of the type described in §3 of Chapter 3. Yet there can surely be no doubt of the result of an actual experiment performed with such clocks.

A further argument due to Dingle [68] is that if 'the asymmetry in the problem arises from the fact that one of the bodies, but not the other, must inevitably be accelerated by some mechanical device' then in Lord Halsbury's problem of three brothers, C's and A's clocks ought to agree on meeting since no accelerations whatsoever occur. This argument is easily refuted by a consideration of the space-time routes of the clocks of B and C, or by M in the original twin problem. In the latter case, the world-line of the 'youthful' route is bent only in the presence of an acceleration; in the former case, an acceleration is unnecessary because non-parallel straight world-line segments are used. These points have been answered in similar vein and in more detail by various writers, especially McCrea. (For details, see the Bibliography.)

We shall now briefly describe some attempts which have been made to introduce accelerated reference frames into special relativity theory, to enable the space-twin as well as the earth-twin to keep running records of the other's progress throughout the whole of the journey.

What is meant by an accelerated reference-frame? Strictly speaking, a reference frame or system is not something associated with an observer in a certain state of motion; it is a scheme for labelling events so that the behaviour of a physical system can be plotted. For example, a particular inertial reference system of coordinates can be used by anyone, whatever his own state of motion, just as anyone listening to news on a radio can use an ordinary atlas to plot world events, even though he may, say, be circling in a jet aircraft at the time. But an inertial observer *can* make use of a reference system particularly adapted to his needs, the inertial reference system in which he is at rest at the origin. One advantage of such a system is that any object, whose space coordinates (x, y, z) are constant, is at a constant distance from him. In a similar manner, an accelerated reference system for a given observer O usually means a scheme of labelling of all events with four numbers or *coordinates* (three of spatial character and one of time character) such that his own space coordinates are $(0, 0, 0)$, and that all objects with constant space

117

coordinates are (in some sense) at a fixed distance from him. This statement is not, however, unambiguous since the measurement of distance involves the simultaneity of observations at separated points, which so far is defined only in inertial systems and is in any case a relative concept.

Various treatments of this problem have been given. Møller [178; 179] introduced a coordinate transformation from an inertial system to one consisting of an infinite succession of rest (i.e. co-moving) systems, and used it in connection with the clock paradox in a case of uniform relativistic acceleration. This treatment assumes the clock hypothesis. Builder [18] has made a detailed analysis including a plot of R's 'world-line' from M's point of view. Another possible approach is by the assumption that the velocity of light is to be constant even for an accelerated observer [153]. This assumption leads to a distance standard. Some writers, including Crampin, McCrea and McNally [40] have taken as a starting point the prescription that M's manner of assigning coordinates is to be as much like R's as possible, i.e. that M mimics a procedure used by inertial observers. Page [185], rather differently, followed some ideas of Milne [284], as employed in the latter's theory of kinematical relativity. A good, though technical account of the different approaches has been given by Romain [197].

We shall consider here, as an example, a simplification of the work of Crampin, McCrea and McNally, and apply it in the case where M is undergoing uniform relativistic acceleration. First, however, we let M be moving in any manner along the x axis of R's inertial frame S (Fig. 23), in which time is denoted by t. We are concerned only with events on the x axis, and therefore only one spatial coordinate will be introduced. At any event E the slope of M's world-line in the diagram is

$$\frac{dt}{dx} = \frac{1}{V}, \tag{4.3}$$

where V is M's instantaneous speed. At the instant that M is at the particular event E he possesses a certain co-moving inertial frame S', in which he is situated at the origin O'. Also shown in Fig. 23 is the simultaneity line EP, through E, in the frame S'; this line has slope V, as follows from the discussion in §8 of Chapter 2.

We suppose that M has set his own clock to read zero at some definite event A on his world-line, and that it subsequently measures proper-time in accordance with the clock hypothesis. How is M to assign coordinates to events? He agrees to give any event such as E, on his world-line, the space coordinate zero and time coordinate T, where T is its time by his clock. Thus, E is the event $(0, T)$ in his own

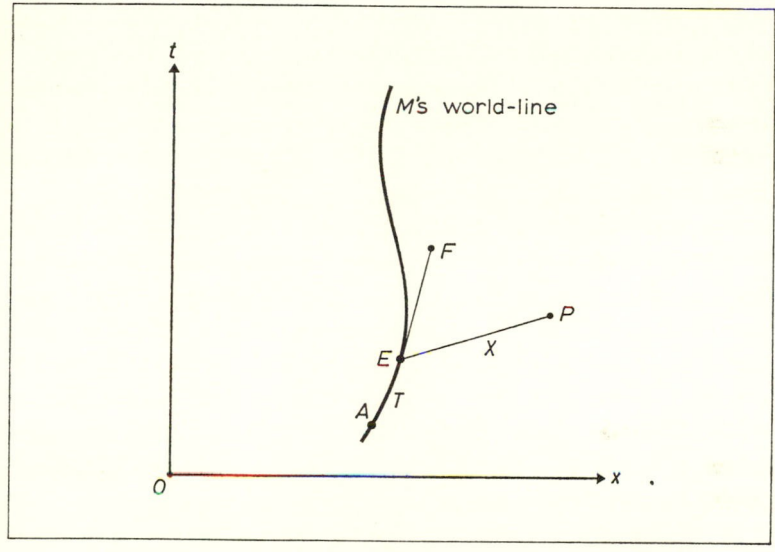

Fig. 23. *EF is tangent to M's world-line at the event E, and EP is the simultaneity line through E in the co-moving inertial frame*

accelerating coordinate system. Any event P on the simultaneity line through E, in the co-moving frame S', is also assigned the time coordinate T, and P is assigned the space coordinate X, where X is the distance EP in S'. This procedure can be said to 'mimic' that appropriate to unaccelerated observers.

It is not obvious that *every* event will be assigned coordinates by M when he adopts this procedure, but Crampin *et al* showed that this is in fact so. In other words they proved that if one considers *any* event P, then there is always an event E on M's world-line such that EP is the simultaneity line in the co-moving frame at E. (The mathematical problem is a nice one. It consists in showing that for any point P in the diagram we can choose a point E on the world-line of M in such a way that EP makes the same angle with Ox (or xO) as the tangent EF makes with Ot.) Crampin *et al* find that there may be more than one such event E, but that the number is always odd! To ensure that M gives an event P only one pair of coordinates (X, T),

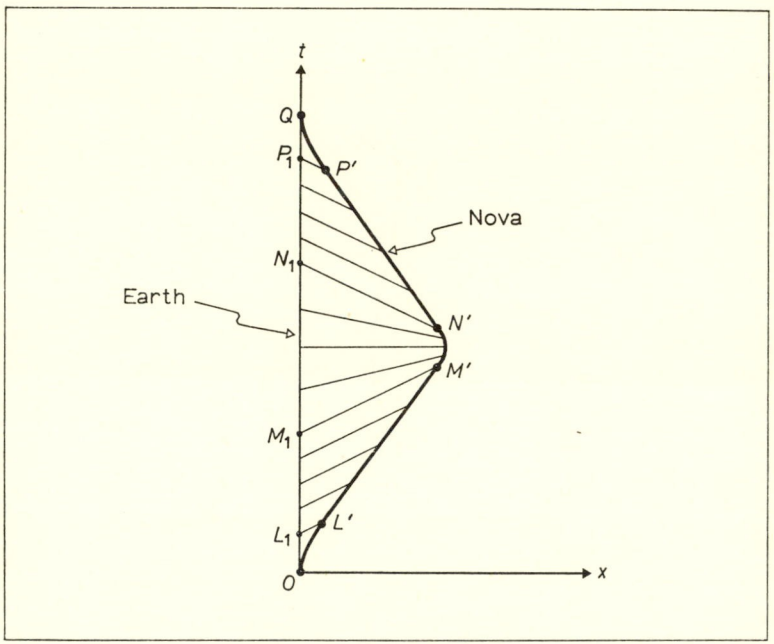

Fig. 24. *Simultaneity lines in accelerated systems for Nova space-ship*

he is instructed to consider only the *first* of the events E when several exist.

The construction allows us to fill in the remaining simultaneity lines in Fig. 17 for observers in the Nova space-ship. They are shown in Fig. 24.

Difficulties in relating the coordinates, assigned by an accelerated observer, to his actual observations limit the usefulness of accelerated reference systems. This is especially true for events very distant from the observer, where the labelling can become extremely erratic through the swinging about of his simultaneity lines. Nevertheless, for the *uniformly* accelerated observer at least, the coordinates can be related to observations.

Let us take the acceleration of M to be of magnitude a in the x direction of S, as described in §3 of Chapter 3. If he starts from the

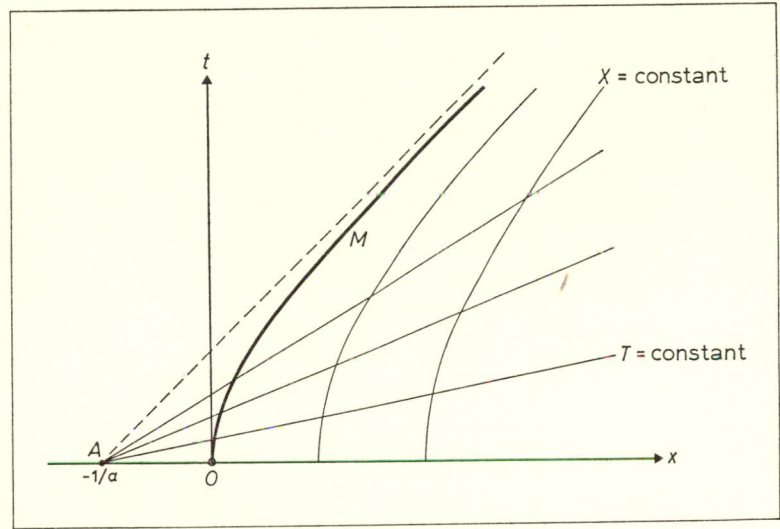

Fig. 25. *Lines of constant X and lines of constant T in the uniformly accelerated system*

origin at $t = 0$, then by (3.5), it follows that his position x is subsequently given by

$$\left(x+\frac{1}{a}\right)^2 - t^2 = \frac{1}{a^2}.$$

The world-line of this motion is shown heavy in Fig. 25. By a direct calculation of his simultaneity lines, or by using some theorems of elementary geometry, it emerges that all these lines pass through the point A ($x = -1/a, t = 0$). Any object, whose space coordinate X assigned by M is constant, has part of a hyperbola as world-line; all such hyperbolae have the same asymptotes through A. Objects with these world-lines all have uniform acceleration in the system S, but the magnitude of the acceleration is different for different values of X. Thus, the curves of constant X are the lines

121

$$\left(x+\frac{1}{a}\right)^2 - t^2 = b^2, \tag{4.4}$$

where $b = X + 1/a$, and $1/b$ is the acceleration of the object at X.

It is interesting to note that an object for which X is constant is at a fixed distance from M in a meaningful sense. In Fig. 26 is shown an outgoing and returning light-signal sent from M and reflected by the

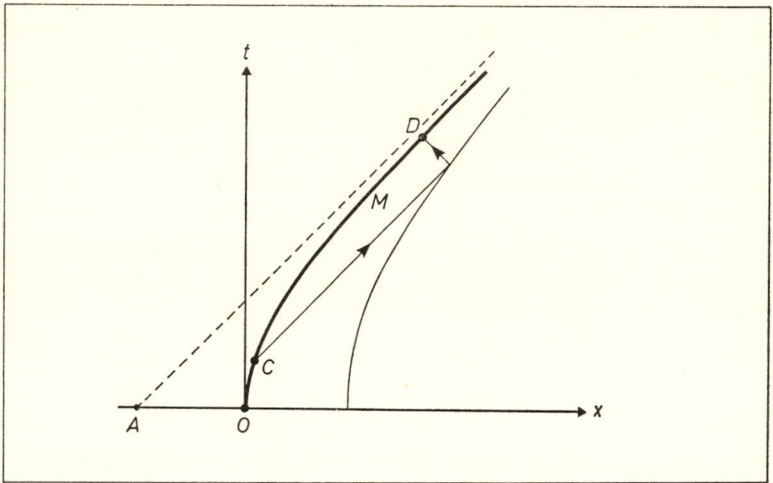

Fig. 26

object in question. The proper-time along the arc CD, where C is the event of emission and D the event of return of the signal, is calculated to be

$$\frac{2}{a}\ln(1+aX), \tag{4.5}$$

which we notice is independent of the particular event C at which the signal is sent. In other words, if the clock hypothesis is valid for M's clock then his radar estimation of the distance of the object is constant. For small values of X, $\ln(1+aX)$ is approximately equal to aX, and (4.5) reduces to $2X$. If the accelerated observer postulates that the speed of light is constant and equal to 1, as for an unaccelerated observer, then he must regard the distance of the body as X.

To sum up; there is no need whatever to introduce an accelerated reference system into the problem to resolve the paradox, but it can give some insight into observations made by an accelerated observer, especially when his acceleration is constant. In more general cases it is difficult to relate M's observations to his adopted coordinates except for events close at hand. This difficulty will be met again in Chapter 6.

2. On matters of simultaneity

Of all disputes over the clock paradox, more have centred on the two issues of acceleration and simultaneity than on all others put together. Objections to deductions of asymmetrical ageing based on the concept of simultaneity (and clock synchronization) are of two main kinds; those which assert that time-at-a-distance is in some way inadmissible, and those in which definite errors are made in its application. In the first kind of objection it is said that the 'rules' are invalid, while in the second the 'rules' are broken. Into the second category fall numerous arguments about clock rates and readings. (See Bibliography.)

Because we have already discussed simultaneity at some length, we shall simply consider briefly one or two of the arguments that have led to more vigorous disputes from time to time. In *Nature*, in 1956, Dingle [56] objected to the use of Einstein's definition of synchronization on the grounds that it was based on an invalid assumption, quoting this passage from the 1905 paper:

'We assume that this definition of synchronization is free from contradictions and possible for any number of points; and that . . . if the clock at A synchronizes with the clock at B and also with the clock at C, the clocks at B and C also synchronize with each other',

and argued:

'We now know that this assumption is only partially valid; two Lorentz transformations in different directions do not commute. Consequently the theory cannot be applied to motion in a polygon and then extended to a closed curve.'

Thus, Einstein's paper 'contains a most regrettable error'; special relativity is powerless to deal with constantly accelerated motions.

Now, in the same year, 1956, Dingle published a paper [59] in the Proceedings of the Physical Society, and himself used Einstein's definition (erroneously) to argue his case for symmetrical ageing in connection with Lord Halsbury's problem. Part of his argument, in our own notation, is as follows. A clock C moves with speed V along

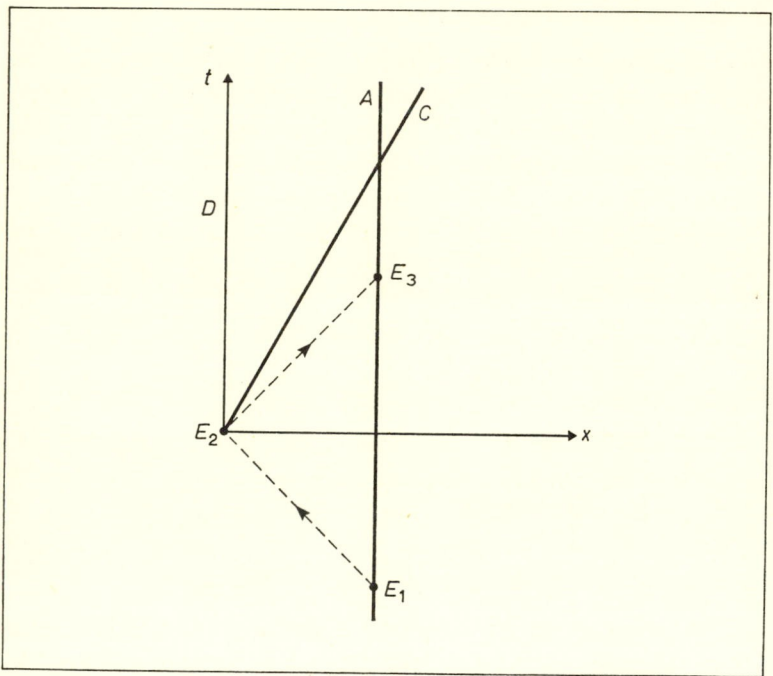

Fig. 27

the x axis of the inertial frame of a clock D. Another clock A is at rest at $x = X$ in D's system. (The various world-lines are shown in Fig. 27.) Clocks C and D are synchronized to read zero as they separate, while A and D are synchronized according to Einstein's definition.

Also shown in Fig. 27 is a light-signal (broken line) sent from A to reach D at time zero, when it is immediately reflected back to A. Events E_1, E_2 and E_3 are those of the emission, reflection and return of this signal. The coordinates (x, t) of events E_1 and E_3 in D's system are respectively:

$$E_1 = (X, -X), \qquad E_3 = (X, X).$$

By applying the Lorentz transformation formulae one finds that in C's frame the coordinates of the same events are:

$$E_1 = (X/u, -X/u), \qquad E_3 = (Xu, Xu) \qquad (4.6)$$

where

$$u = \sqrt{\frac{1-V}{1+V}}.$$

Dingle applied the simultaneity definition to the time coordinates in (4.6), to deduce that the time of E_2 in C's system is $\frac{1}{2}(Xu - X/u)$, which is less than zero. Since, however, the clock C is known to read zero at this event, Dingle's conclusion is that C 'reads fast' and that an approaching clock would likewise read slow (by Einstein's definition), and that these anomalies are the source of misconceptions in others' arguments.

Replies to Dingle on this kind of issue have been made by McCrea, Crawford and others. Crawford [42] argues that it is wrong to apply the definition in order to synchronize a local stationary clock with a distant moving clock, and that Einstein himself only used the definition for clocks relatively at rest. Since the definition assigns a time to any distant event, it is, in fact, possible to synchronize a clock in any state of motion with a local stationary one belonging to the observer in question, though naturally the synchronization will not be maintained. The precise error in Dingle's procedure is that E_1 and E_3 are at different points in C's frame; the signal is sent from one point and returned to another, and so the procedure cannot be expected to produce useful results.

L. Essen [91] agreed with Dingle that Einstein had made an error, and explained that in discussing a moving clock, C, one must distinguish between the impulses of its standard and the readings on its dial, because the latter will automatically relate to measurements

125

made on C. (But if a pointer on C's dial shows the reading n as the nth impulse occurs, it is hard to see how any observer can find a true distinction between the 'impulse time' and the 'dial time' of C.) Later [95], Essen maintained: 'It has been shown that the well-known clock paradox is predicted by the theory only by changing the meaning of the symbols during the course of the argument.' Apparently this was in reference to the use of one symbol for both the impulse and dial times.

In a thoughtful article, H. E. Ives [132] discussed to what extent special relativity theory can be developed without appeal to Einstein's definition. He made use of two-way light signals in timing distant events, but, in keeping with some ideas of A. A. Robb (developed in Robb's book *Geometry of Space and Time* [291]), assumed that the time at which reflection occurs is indeterminate except that it is later than the time of emission of the signal and earlier than the time of return. This assumption is equivalent to the statement that the one-way speed of light might have any value which is finite and non-zero (only the average of the outward and return speeds being invariable in his treatment).

Others have sought to develop special relativity entirely on the basis of 'equivalent' observers on the lines pioneered by Milne in his theory of kinematical relativity. This ingenious theory starts from the hypothesis that at each point of space there is (at any instant) *one* state of motion that is fundamentally distinguished from all others, rather than a whole family of motions (the inertial motions) as in special relativity. The physical basis of this hypothesis is that in the universe each cluster of galaxies (roughly speaking) has a definite motion and the motion is presumed to be such that the overall view of the universe, as the universe expands, is the same from each cluster. (This is the so-called 'cosmological principle'; see, for example H. Bondi, *Cosmology* [247].) In particular, the timekeeping ideas together with the notion of equivalence enabled Milne to derive the Lorentz transformation between the coordinate systems of any pair of equivalent observers and so construct a kinematical theory of relativity [284].

Milne's ideas have been followed by Whitrow, Page [185] and various others in connection with time-at-a-distance and the synchronization of clocks. We shall give here by way of illustration a brief account of Whitrow's approach.*

Whitrow starts with almost a blank sheet. An observer A is equipped with a suitable standard clock and sends a signal (*not necessarily a light-signal*) to an observer or mechanical device at B where it is immediately reflected back to A. At this stage nothing is

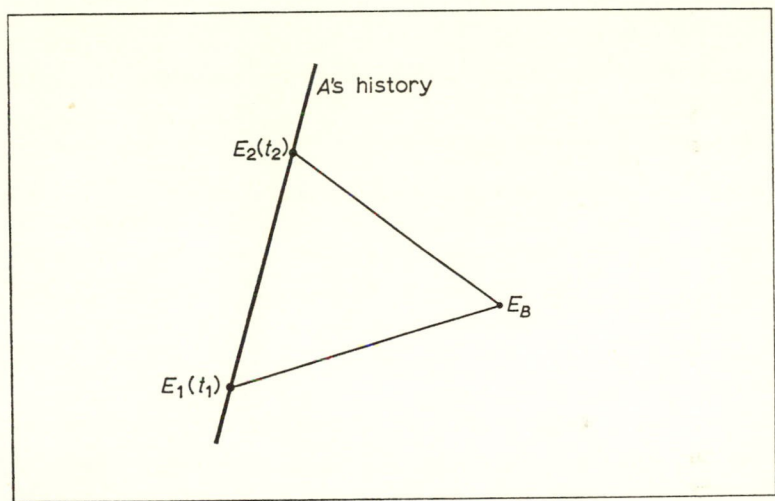

Fig. 28

said about the motion of B, and the object is to decide what time A should assign to the event, E_B, of the reflection at B. Let E_1 and E_2 be the events on A's history of the emission and return of the signal; and let his clock times of these events be t_1 and t_2 respectively. The situation is shown schematically in Fig. 28 (which however is not a space-time diagram because distance and time coordinates have not yet been defined).

Whitrow assumes that of all possible signals that might be used there is one type that is the fastest. By this statement it is meant that

*An abridgement from his book *The Natural Philosophy of Time*, Nelson, London and Edinburgh (1961). I am grateful to Dr Whitrow and his publishers for permission to include it. (Copyright: G. J. Whitrow, 1961.)

for a particular event E_B there is one event E_1 which occurs at the latest possible time on A's history at which a signal can be emitted so as to arrive at E_B. Similarly, E_2 is the earliest event on A's history at which a signal from E_B can arrive. (Clearly one has in mind light-signals, though particular properties of light propagation are not assumed.) Next, are introduced a sequence of axioms.

Axiom 1. Postulate of causality: t_2 exceeds t_1 unless E_B occurs at A, in which case $t_2 = t_1$.

This modest axiom simply means that signals do not return before they are sent.

Axiom 2. Postulate of spatial isotropy: the epoch (time) t_B which A assigns to E_B is determined by a relation of the form $t_B = f(t_2, t_1)$, where f is a single-valued function of t_2 and t_1.

For example, in Einstein's definition $f(t_2, t_1)$ is identically $\frac{1}{2}(t_2 + t_1)$. Axiom 2 implies that the time assigned to E_B is not to depend on the direction in which the signal has to be sent in order to arrive at B.

For convenience, fastest signals are given the name *light-signals*, and the notion of time ordering of all events on the path of a light-signal is introduced. Whitrow also assumes that the light paths joining pairs of events on them are unique:

Axiom 3. The light paths joining E_1 with E_B and E_B with E_2 are, in general, unique.

Let E_C be an event after E_B on the light path $E_1 E_B$. (We may suppose that E_C is an event at some mechanism or observer with position C.) In Fig. 29, a light-signal from E_1 to E_C, via E_B, is immediately returned towards A. It arrives back at A at event E_3, when A's clock reads t_3. By axiom 2, A assigns the time t_C to event E_C, where $t_C = f(t_3, t_1)$. Evidently t_3 is greater than t_2, since the direct light-signal is the fastest signal from E_B to A. As a fourth axiom the condition is imposed that A's time ordering of the events E_B and E_C agrees with the order they occur on the light path. That is to say, t_C is greater than t_B in the case depicted in the diagram. This rather mild assumption imposes quite a strong restriction on the form of the function f, namely, that $f(t_3, t_1)$ must be greater than $f(t_2, t_1)$

whenever t_3 is greater than t_2. The object, now, is to introduce further axioms in order to determine f more fully.

In the next stage, A is instructed how to assign distances between events like E_B and E_C on the light path. The distance is to be a positive quantity which depends on t_B and t_C only; (this condition

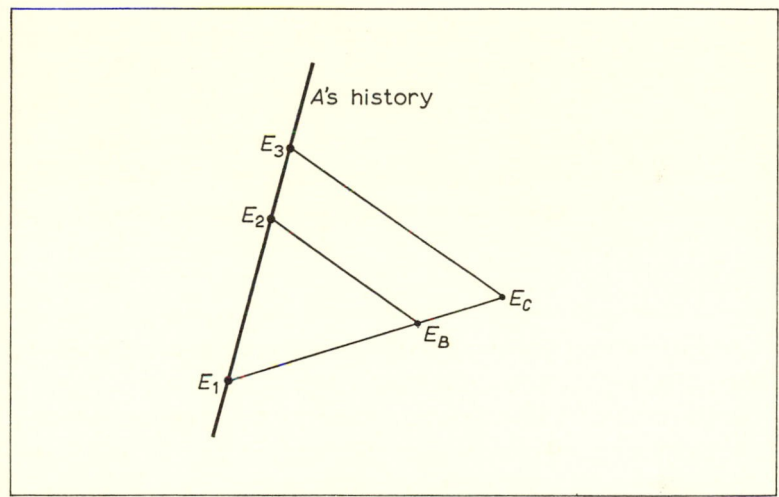

Fig. 29

involves a certain assumption regarding the homogeneous nature of space). Further axioms impose requirements on the distance formula; for example, the addition law

$$\text{dist. } (E_C, E_B) + \text{dist. } (E_B, E_1) = \text{dist. } (E_C, E_1)$$

is introduced; and also the requirement that

$$\text{dist. } (E_1, E_B) = \text{dist. } (E_B, E_2),$$

which expresses A's view that he himself is at rest (i.e. at a fixed distance from the event E_B). After seven axioms in all, the stage is reached where the distance r of the event E_B from A is given by an equation of the form

$$r = \tfrac{1}{2}[g(t_2) - g(t_1)], \tag{4.7}$$

129

and the time t assigned to E_B must satisfy

$$g(t) = \tfrac{1}{2}[g(t_2)+g(t_1)] \qquad (4.8)$$

in which g may be any single-valued strictly increasing function. On eliminating $g(t_2)$ between the last two equations we get

$$r = g(t)-g(t_1).$$

By fixing the event E_1 (and hence fixing t_1) we deduce that A's measure of the speed of an outgoing light-signal is $dr/dt = dg/dt$. This is also the speed of an incoming light-signal, as follows by eliminating $g(t_1)$ instead of $g(t_2)$ between (4.7) and (4.8). Einstein's definition of time-at-a-distance, as implied in his clock synchronization, corresponds to the case $dg/dt = c$, when (4.7) and (4.8) reduce to

$$r = \tfrac{1}{2}c(t_2-t_1),$$
$$t = \tfrac{1}{2}(t_2+t_1).$$

Whitrow shows that these final forms are a consequence of two further axioms. These are that the resetting of A's clock (by changing his *zero-point*) does not affect his measure of time-intervals, and that if he changes his *unit* of clock time then the times he assigns to distant events are changed by the same constant factor.

We can thus see one form of axiomatic basis for Einstein's particular synchronization procedure. For details of further developments of this interesting approach, involving the concept of the equivalence of several observers, the reader might consult Whitrow's book or Milne's *Kinematic Relativity*.

3. *The Doppler effect and the k-calculus*

The Doppler effect is the observed change in frequency of a wave, or other periodic signal, when the motion of the source or the observer is changed. It has been used in a number of discussions of the twin and three-brother problems in order to eliminate time-at-a-distance. Particularly illuminating is the argument based on the k-calculus (a calculus generally attributed to H. Bondi, who used it in a series of broadcast lectures). The Doppler effect formulae are obtained very simply by means of the k-calculus; alternatively, the Lorentz transformation may be used directly for this purpose. Furthermore, the k-calculus may even be used to *derive* the Lorentz transformation (see D. Bohm, *The Special Theory of Relativity* [246].

In an exposition in *Discovery* Bondi [11] considers three inertial observers, whom we shall call A, B and C. They are situated in a straight line with A and C at relative rest, and B is between them and

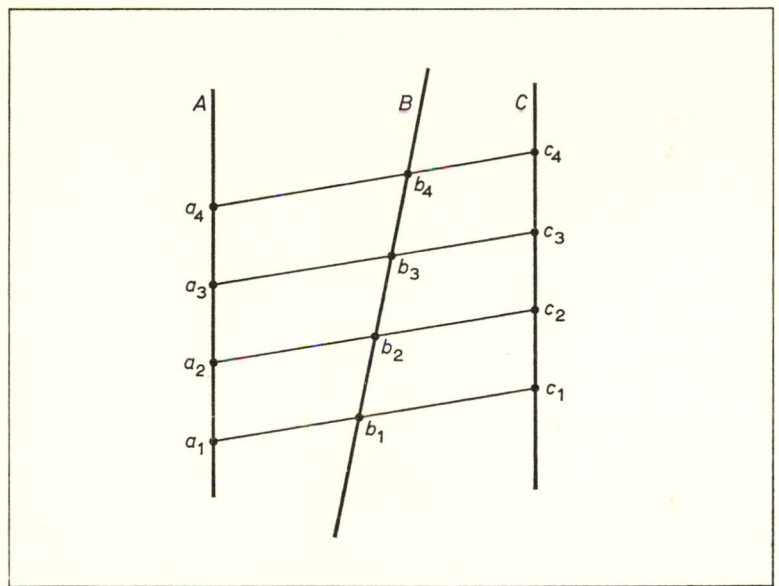

Fig. 30

moving with speed V in the direction AC (Fig. 30). A emits a sequence of light signals (or radio pulses) at one second time-intervals, the events of emission being $a_1, a_2, a_3, \ldots$, and these signals pass B at events $b_1, b_2, b_3, \ldots$ and arrive at C at events $c_1, c_2, c_3, \ldots$.

In the frame of A and C, the signals take equal times to travel from A to C, and so the time-intervals between the receptions of successive signals are the same, according to C, as the time-intervals between emissions according to A. These time-intervals are all one second.

Since B is receding from A, successive signals take progressively longer to travel from A to B, but they evidently arrive at what B regards as a uniform rate. Let B's measure of the time-interval between successive arrivals be k seconds, where k will depend on V.

It is clear from the diagram that had B himself chosen to send light signals to C, at precisely the events $b_1, b_2, \ldots$, then these signals would also have arrived at C at the events $c_1, c_2, \ldots$. In other words, C's interval of reception is $1/k$ times B's interval of emission.

We therefore have the important point that if k is the ratio of reception interval to emission interval when two observers are mutually receding at speed V, then $1/k$ is the corresponding ratio when the observers are mutually approaching at speed V.

The factor k is easily calculated, and it is also known experimentally. But the power of the k-calculus approach to the twin problem is that assumptions are minimal if one accepts only the easily observed fact that k is not equal to 1. (The observed 'reddening' of light from a receding source shows in fact that k is *greater* than 1.) In Fig. 31, the traveller M separates from the inertial twin R at speed V, and after a brief period of acceleration returns to R at the same speed V. E_1, E_2 and E_3 are the events of separation, reversal and return, respectively. It is supposed that M sends signals at unit time-intervals (by his own clock) throughout the journey.

Let t_1 be M's measure of the times taken to travel between events E_1 and E_2, and also between E_2 and E_3. Let 0, t_0 and T be the times by R's clock for the events of separation, receipt of M's signal from E_2, and return. Then R receives 'slow' signals at intervals k for a time t_0, and 'fast' signals at intervals $1/k$ for a time $T - t_0$. If we equate the total number of signals received to the number of signals sent, for each part of the journey, we get

$$t_0/k = t_1, \qquad (T - t_0)k = t_1, \tag{4.9}$$

and on eliminating t_0,

$$T = (k + 1/k)t_1. \tag{4.10}$$

But $k + 1/k$ is always greater than 2 for any positive value of k other than $k = 1$, and so M's recorded time $2t_1$ is less than R's recorded time T, in agreement with the usual asymmetrical result. A similar argument with signals sent from R to M does not contradict this

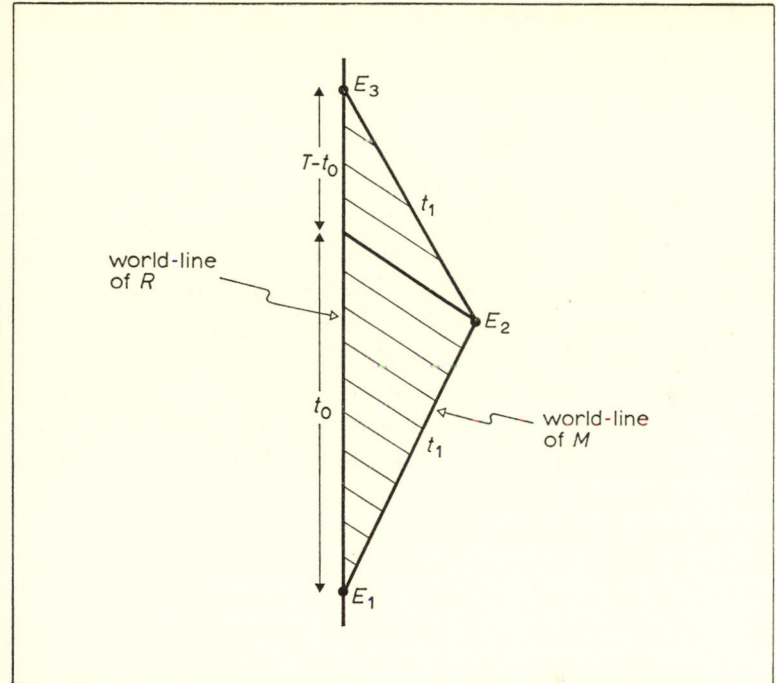

Fig. 31

result, because M would then receive more fast signals than slow ones.

To evaluate k, we first eliminate t_1 from equations (4.9) and write

$$t_0/k = (T-t_0)k. \tag{4.11}$$

Next, time-at-a-distance has to be introduced. According to R, M travels away for a time $\frac{1}{2}T$, when he reaches a maximum distance $\frac{1}{2}TV$, and therefore the signal arriving at time t_0 has travelled (at speed 1) for a time $\frac{1}{2}TV$. Hence,

$$t_0 = \tfrac{1}{2}T + \tfrac{1}{2}TV,$$

and if this value is substituted for t_0 in (4.11), we find on rearranging that

$$k = \sqrt{\frac{1+V}{1-V}}. \tag{4.12}$$

If β denotes the quantity $1/k$,

$$\beta = \sqrt{\frac{1-V}{1+V}}, \tag{4.13}$$

then this is the ratio of the *frequency* of reception of signals by R to the *frequency* of their emission by M, when M is receding with speed V. The formula (4.13) for the ratio of frequencies is the standard relativistic formula of Doppler effect in the case of direct mutual recession of source and receiver. The case of mutual approach at speed V is similar, with V replaced by $-V$ in (4.13).

J. H. Fremlin [106] gave a Doppler effect argument (not using the k-calculus) to demonstrate asymmetrical ageing, in an article in *Nature* in 1957. His twins 'Traveller' and 'Stay-at-home' exchange radio signals from identical crystal-controlled short-wave transmitters. Darwin [47] gave an equivalent simple discussion, denoting Traveller by S_1 and Stay-at-home by S_0. Each emits light-signals at the rate n per unit time, and the average rates of reception of these are calculated. Both writers obtain results which exhibit asymmetry. Generalizing Darwin's discussion, one finds that S_0 receives $n\beta t_0$ 'slow' signals and $(T_0 - t_0)n/\beta$ 'fast' ones, where T_0 has the same meaning at T in the above discussion. Similarly, S_1 receives $n\beta t_1$ 'slow' and $(T_1 - t_0)n/\beta$ 'fast' signals, where T_1 is S_1's total time for the journey. Equating the total number of signals received to the total number emitted, in each case, we get

$$\begin{aligned} n\beta t_0 + (T_0 - t_0)n\beta &= nT_1, \\ n\beta t_1 + (T_1 - t_1)n/\beta &= nT_0, \end{aligned} \tag{4.14}$$

Dingle [66] disputes the interpretation of these equations as indicating asymmetry, in that T_1 must be less than T_0, although if t_1 is put equal to $\frac{1}{2}T_1$, the second equation gives

$$\tfrac{1}{2}(\beta + 1/\beta)T_1 = T_0,$$

which demonstrates it at once. Dingle is here replying to Darwin (and refers also to Fremlin) but his case is weak. He produces equations (4.14) as 'common ground', and proceeds from there under the assumption of symmetry, thus setting $t_0 = t_1$. But (4.14) are not consistent with this if $t_1 = \frac{1}{2}T_1$, and so something must give way.

Dingle's explanation is that t_1 is greater than $\frac{1}{2}T_1$, which leads him to assert that 'S_1 will not observe S_0's flashes to change [frequency] until *after* he has fired his rocket'. Dingle argues, therefore, that S_1 observes no *immediate* change, after firing his rocket, in the rate at which he encounters signals from S_0. The change comes later on. Darwin's postulates he regards as equivalent to the introduction of a fixed ether against which velocities are measurable, and that to locate the reversal at S_1 uniquely is not compatible with the relativity viewpoint.

It would be of great interest to find experimental evidence of such a delayed Doppler effect. Fahy [100] has made the point that it could easily be sought in astronomy. In the earth's orbital motion, its velocity of recession (or in a few cases, approach) relative to the distant nebulae is constantly changing, giving rise to a change in the colour of light received from them. The velocity change should result in the oscillation of any visible spectral line about its mean position with a period of oscillation of one year, and a maximum change in wavelength of about 1 part in 10^4. Due to Dingle's delay effect, however, these oscillations should be out of phase with fluctuations of the earth's component of velocity towards the particular nebula by a number of years. In general, the delay will not be a whole number of years (it is equal to the time the light takes to reach us) and will be different for different nebulae. Thus, observations of a number of nebulae should reveal oscillations of the spectral lines with period one year but with randomly differing phases. Apparently no such observations have been made; it would be most remarkable if they had.

4. When is a clock not a clock?

We have given instances of what *does* constitute a valid clock for use in inertial frames and for use by accelerated observers. One might conversely ask, in certain cases, what does *not* constitute a valid clock; or more simply, when is a clock not a clock? The question became especially relevant to the paradox controversy in 1939 when Dingle [50] described timepieces which apparently were not subject to the time-dilatation phenomenon predicted by special relativity.

These devices were introduced in an article, entitled 'The Relativity of Time', wherein he maintained that while (circumstantial) evidence could be found for the contraction of a moving rod (in the Michelson–Morley and other experiments), there was not, and could not be, evidence for the slow running of moving clocks, because in physics there is no explicit definition of a clock. He gave the example of an hour-glass in which grains of sand fall at the 'standard rate' into a container, the time of an occurrence being measured by the number, N, of grains which fall between an arbitrarily chosen zero instant and the occurrence in question. Let us suppose that the hour-glass clock is at rest in a certain inertial frame and that it is used to measure times of events close to the clock. If M and V are respectively the mass and volume of a particle of the sand used, then the mass and volume of N such particles are respectively MN and VN, and each of these is proportional to N. Therefore, both the mass of sand fallen and the volume of sand fallen can also be used to measure time (a suitable constant factor being introduced as necessary to accord with any choice of time unit).

Dingle supposed the clock to be set in motion with constant speed v, and that 'when it is moving uniformly, N is changed to N'. (The precise meaning of this statement may be left open for the moment.) Then MN changes to $MN'/\sqrt{(1-v^2)}$ (by the mass-velocity law) and VN changes to $VN'\sqrt{(1-v^2)}$ (by Fitzgerald-Lorentz contraction). Dingle's case is that if an observer wishes to compare the rate of the moving hour-glass clock with his own stationary one, then he might equally well measure rates by the number, mass or volume of the grains fallen. The ratios of rates of the clocks in the three cases are (with β denoting $1/\sqrt{(1-v^2)}$),

$$\frac{N'}{N}, \quad \frac{N'\beta}{N}, \quad \frac{N'}{N\beta}, \tag{4.15}$$

and the inequality of these ratios is taken to imply that 'the time-coordinates of the [Lorentz transformation] formulae does not represent (except possibly accidentally in a particular case) the reading of a clock'. He claims that t is the reading of a clock at rest, but t' is not in general the reading of a clock in motion. In particular, t is to be measured by the frictionless passage of a free particle along a graduated rod in inertial motion, in accordance with Newton's first law (which identifies t with ephemeris time) and that an 'ideal' clock at rest would measure this variable t. This led Dingle to affirm that t' must be measured in the same way and that while a moving *ideal* clock runs slow, other types need not do so. The Kennedy-Thorndike experiment, he suggested, was not a confirmation of the Lorentz transformation for time, but an indication that the frequency of a radiating atom behaves as an ideal clock.

The mass and volume clocks have been criticized by Campbell [27] who maintains that it is not legitimate to measure, for example, the mass of sand fallen at any stage in the clock belonging to one system in terms of the unit of mass in the other system, it being necessary to use the Lorentz equations to relate the mass units.

In his reply to Campbell, Dingle [51] stressed his intention to remove a general notion that measuring time is identical with reading the dial of an unspecified instrument, and drew an analogue with different types of thermometer (corrected mercury-in-glass, platinum resistance, etc.) which do not measure absolute temperature, just as clocks need not measure physical time. One should not infer that 'anything that ticks and moves a pointer round a dial will visually display the Lorentz transformation'.

Later he referred [52] to the quantity of correspondence that he had received on the matter. Some correspondents found it important to introduce a second observer, to move with the moving clock. But if this requirement means that rest-mass or rest-volume is to be used (so that all three of the ratios (4.15) reduce to N'/N) then the two clocks would always agree over time-intervals, he argued, because *number* is the same for all observers. (This shows the looseness – not uncommon at that time – of statements about time measurements. The meaning of the ratio N'/N is vague. Two observers will not dispute which two events, at a particular clock, are those of the falling of the 1st grain and the Nth grain of sand. But they will not necessarily agree as to which events, at another clock, are simultaneous with these.)

One notable exchange on this issue was between Dingle and Epstein [89] (in the *American Journal of Physics*). The latter stressed the operational viewpoint according to which a physical concept (such as time-interval) is not defined until one prescribes an instrumental operation which reduces it to measurements. His criteria for legitimate clocks were based on the interpretation of clock readings as events, and hence the mass and volume 'constructions' have 'hardly more relevance to time measurement than the famous attempt of Sir Isaac Newton, who, wishing to boil a four-minute egg, reputedly, put his watch into the boiling water and was found by his housekeeper staring at the egg in his hand'.

Dingle [53] later objected to Epstein's insistence that *two* observers are involved in the problem and reaffirmed his belief that the number clock would not exhibit time-dilatation. Sorely tried, Epstein [90] complained, in a final report, of 'gratuitous buffoonery' and the 'burlesquing' of his own analysis of the Lorentz transformation formulae.

The eminent Polish physicist, Leopold Infeld, asked to adjudicate by the editor of the *American Journal of Physics,* produced an account [123] of the properties of 'periodic' and 'aperiodic' clocks. Periodic clocks include watches, ordinary clocks, an hour-glass turned over whenever the upper container is empty, etc., while aperiodic clocks are typified by the example of a free particle along a straight graduated scale. Infeld considered the number clock to be essentially periodic, and the mass and volume clocks to be aperiodic, and argued that the apparent rates of aperiodic clocks in relative motion could not be unambiguously compared. This account of time measurement is interesting, though I think that its importance in the present context was emphasized unduly by Infeld.

The central issue is surely the comparison of two identical clocks by an observer located at one of them. The observed slowness of the rate of the other clock is predicted, according to the Lorentz transformation, for clocks of all three types, number, mass and volume, provided that each clock measures the time variable (t or t' as the case may be) in its rest frame. The proof follows the argument given in §7 of Chapter 2, and is clearly concerned with time-intervals between certain *events*, whether these be 'ticks', dial readings, or the falling of the Nth grain, gramme or millilitre of sand. In the latter two examples, the reason that only the *rest-mass* or *rest-volume* is to be used is that the observer and the dial (or other indicator) of the moving clock must agree as to the event at which the Nth gramme or millilitre has fallen.

5. The principle of impotence

'Were a new Isaac Newton born today, we could send him space-travelling so as to return to us in thirty years time at, say, the age of three. This would be in accordance with the theory of relativity and with experimental tests of the theory – and all we should get for our efforts would be a retarded child.'

McCrea's observation [166], in 1957 (surely one of the bumper years for writers on the clock paradox), was made as a consequence of a remark to him by J. S. Courtney-Pratt that if asymmetric ageing is possible a space-traveller could get his problems solved more quickly by getting an earth-dweller to do the thinking. McCrea sought, perhaps lightheartedly, to reassure people who would deny asymmetrical ageing because it appears to give something (i.e. longer life) for nothing, whereas nature never works that way. Such people are thus convinced that there must be a principle of nature that disallows this kind of profit.

A large number of laws of nature can be expressed in the form of general principles of prohibition. Whittaker [302] listed several in a lecture some years ago. Examples include:

'It is impossible to set up an electric field in any region of space by enclosing the space in a hollow conductor of any shape or size and charging the outside of the conductor:

'It is impossible to derive mechanical effort from any portion of matter by cooling it below the temperature of the coldest of the surrounding objects.

'It is impossible to detect a uniform translatory motion, which is possessed by a system as a whole, by observations of phenomena taking place wholly within the system.

'It is impossible to tell where one is in the universe.'

The first of Whittaker's examples contains a significant part of the foundations of electromagnetic theory. The second is a consequence

139

of the second law of thermodynamics. The third is a form of wording of the principle of special relativity, and the last is a statement of the cosmological principle (not the *perfect* cosmological principle; see H. Bondi, *Cosmology* [247]) which is assumed in almost every cosmological theory, and which means that the universe presents the same overall aspect at all points. Many similar principles of nature can be listed; Whittaker summed them up with the name *postulates of impotence.*

As McCrea pointed out, if asymmetrical ageing were to work the other way round (so that moving clocks ran *fast*) then we could get Newton II back after three years at the scientifically productive age of thirty (ignoring, for the moment, technological and other limitations on space-travel). But as things are, we can only *lose* by the experiment.

In a similar way, we are prohibited from increasing the effective speed of our best computers by sending them into space while they perform lengthy calculations, again because of the direction of asymmetry. McCrea's principle of impotence may be expressed as thus; in a freely moving laboratory it is impossible to put ourselves at an advantage, in obtaining physical results, by making use of the different passage of time in another laboratory constrained to move in any other way.

Nevertheless, time-dilatation can, in principle, be used to advantage. An individual could, for example, take a short vacation in space while his laboratory computer toiled for years on his calculations. Again, the incurably ill might take such vacations to await the discovery of a cure on earth. This seems far-fetched, but it *might* one day be possible, and could well be preferable to the artificial hibernation of the incurably ill (by deep-freezing) as some have seriously suggested. (For problems relevant to the latter, see A. U. Smith *Biological Effects of Freezing and Supercooling* [298].)

6. Special relativity: right or wrong?

The validity of the special theory has been questioned in numerous discussions of time-dilatation. On a number of occasions H. E. Ives examined the implications of various optical experiments and proposed the rejection of their relativistic interpretation in favour of the hypothesis of a luminiferous ether. In 1937, for example, he based his case [126] on the evidence of stellar aberration, and concluded that time is absolute, while introducing the concept of clocks 'unaffected by transport'. Though believing that distant simultaneity was meaningful, he took it to be indeterminate within certain bounds (in the manner of Robb; see §2), and on this basis obtained transformation formulae akin to those of the Lorentz transformation.

One argument due to Ives [125] was that the Michelson–Morley experiment could be explained by contractions of the apparatus in both the direction of its motion through the ether (at speed V) *and* in the transverse direction, provided that the two contractions were in the ratio $\sqrt{(1 - V^2)}:1$. This, of course, includes the hypothesis of Fitzgerald and Lorentz as a special case. In connection with the clock paradox, Ives argued [127] that it presents no difficulty if the ether is the framework with respect to which the alteration in clock rates takes place. However, he appeared to consider only a choice between the ether argument and one in which relative velocities alone are taken into account, thus omitting the conventional relativist's view.

Ives [128] also attacked the arbitrariness of operational procedures involved in Einstein's theory, saying:

'By virtue of its actual, though unacknowledged, derivation from the ether and the contractions of length and frequency which occur on motion relative to the ether, it predicts correctly the results of performing *certain selected operations*. Neglecting to inquire into what physical behaviours underly the validity of this special case, *it offers no explanation of what lies behind the particular choice of operations*. It thus gives only a partial account of the phenomenon of moving bodies, in contrast to the complete account given by the hypotheses of Fitzgerald, Larmor and Lorentz, based on the phenomena of aberration and the Michelson–Morley experiment.'

The Kennedy–Thorndike experiment (using a Michelson–Morley type interferometer, with unequal arms) showed the relativity of time because the assumption was made that there was a contraction of length in the direction of motion. But had they made other assumptions, he argued, they could equally have proved the *non*-relativity of

141

time. Ives [130] expressed the belief, however, that the experiment performed by himself and G. R. Stilwell on the transverse Doppler effect (see Chapter 5), was evidence for the particular contraction and dilatation phenomena in special relativity. And in 1951 he gave an account [133] of the clock paradox in which the Lorentz transformation was used with no reference to the ether.

Geoffrey Builder was another who argued a case for the ether, and while much of his work was of undoubtedly high quality, it seems that this aspect of it was based on an initial misconception. In a discussion [18] published in 1957, he convincingly argued that the clock paradox could be resolved in terms of special relativity. (This led to a lively dispute with Dingle [60; 19; 61].) But the following year he appeared to have a change of heart when he reasoned as follows [21]. Let A and B be standard clocks which separate at time $t = t_1$ and reunite at time $t = t_2$ in an inertial frame S. Let u and v be the speeds of the respective clocks in S, where u and v depend on the time t. If clocks A and B record total times T_A and T_B, respectively, during the interval between separation and reunion, then according to the clock hypothesis these recorded times will differ by an amount

$$T_A - T_B = \int_{t_1}^{t_2} \sqrt{(1-u^2)}\, dt - \int_{t_1}^{t_2} \sqrt{(1-v^2)}\, dt, \qquad (4.16)$$

(and such a formula will apply for every choice of the inertial frame S). Builder insisted that since the retardation of one clock relative to the other, in this formula, is not a function of the *accelerations* of the clocks, this precludes 'the possibility of a causal relation between relative retardation and these accelerations'. It is hard to see any justification for his claim, however, because speed and acceleration are so inextricably related. In the complete absence of accelerations the world-lines of the clocks would be coincident straight lines in space-time, u and v would be identical in (4.16), and therefore no relative retardation would occur. Thus, the role of acceleration in (4.16) is indisputable.

A strict analogue of this situation concerns the length of a curve between two points in the Euclidean plane. Suppose that these are the points $(0, 0)$ and $(1, 0)$ in a certain cartesian coordinate system; then the arc length of a curve $y = y(x)$ joining them is given by

$$\int_0^1 \sqrt{(1+y'^2)}\, dx, \qquad (y' = dy/dx).$$

Can one reasonably argue that because the arc length is seen to depend only on the slope y' it is in no way determined by, say, the curvature of the curve, as though slope and curvature were not interdependent? Yet this is the type of argument Builder seemed to rely on.

However, it should be emphasized that Builder's work is of interest because it contains a penetrating analysis of the compatibility of the ether hypothesis and the theory of relativity.

The experimentalist Essen [91] believed that the clock paradox arose through an error in Einstein's presentation which had not hitherto been noted and which, he thought, might be more obvious to an experimental rather than a theoretical physicist. The 'error' concerned the lack of distinction between the 'impulses' of a time standard and the readings on its dial, a distinction to which we have referred in §2. Essen, too, asserted that since no data concerning the effects of acceleration are included in the theory of clock retardation, the result cannot be a consequence of acceleration. The theory of relativity thus stood in peril, for if the views he presented were correct,

'the consequences are far-reaching. Many text-books need to be revised, and as the paradox seems to be confirmed by the general theory of relativity some aspects of this theory may need to be critically examined.'

(*Had* the views been correct, this would, of course, have been true.)

Dingle, on the other hand, in his reference (in 1956) to the 'regrettable error' in Einstein's 1905 paper, did not suggest that this invalidated the special theory itself [56]. His doubts on the latter came into evidence later on. In 1962, he referred [72] to an argument he had put forward in support of 'symmetry', and concerning the apparent incompatibility of the equations $\tau = t\sqrt{(1-V^2)}$ and $t = \tau\sqrt{(1-V^2)}$. (Here, t and τ were to be interpreted as elapsed times since a common zero event, recorded by two clocks respectively at rest in inertial frames k (with coordinate system (ξ, τ)) and K (with coordinate system (x, t)), the relative speed being V. (In fact, the meaningful

143

interpretation of the two formulae is that t is a coordinate and not the reading of a single clock in the first formula, and that τ is a coordinate and not the reading of a single clock in the second. The equations are therefore not incompatible since the symbols have different significance in the two cases.) Now, however, Dingle interpreted the supposed incompatibility as a genuine contradiction.

Worried by the possible consequences of the continuing reliance of scientists on a theory shown to be false, he expressed much concern that the argument had produced no response from others. The late German scientist Max Born [13] reported (in 1963) receiving a reprint of Dingle's 1962 paper with the handwritten comment: 'With kindest regards. Test case for the integrity of scientists.' In a sense, it was a pity that Born then took up the challenge, because a satisfactory reply to Dingle needed more time than Born wished to devote to the matter. His brief reply, in *Nature*, consisted largely of a 'correction' to Dingle's question (hardly likely to produce the desired effect) and a partially explained space-time diagram.

At about this time, Essen [94] produced what must be one of the most remarkable of all suggestions against the 'conventionalist' view, in connection with the Doppler effect type of discussion (§3). According to Essen, 'the assumption that all the timing pulses sent out by the moving clock are received by the stationary observer during the time of the journey' is made by H. Bondi, Darwin, etc., and is the only way the result can be derived from the special theory. '*It is easy to show that there are both practical and theoretical reasons why this assumption cannot be made.*' [My italics]. What happens to the signals that do *not* arrive is, however, unexplained.

The validity of the special theory in more recent years (in so far as the issue of time is concerned) has been argued afresh by McCrea and Dingle, whose detailed cases 'for' and 'against' in *Nature* (in 1967 and 1968) have become well known [75; 76; 167]. Many analyses of the premises and logic of relativity theory have been made since 1905, but none appears to have evoked such vigorous exchanges as the ones centred on the phenomenon of time-dilatation. And although physicists generally are convinced that the special theory is consistent and valid, these exchanges will no doubt continue.

Chapter 5

THE EXPERIMENTAL EVIDENCE

'What is now proved was once only imagined.'

WILLIAM BLAKE, *Proverbs of Hell*

1. Before 1950

Apart from indirect evidence, such as that provided by the Michelson–Morley experiment itself, there were three main sources of evidence of time-dilatation until about 1950. These were the Kennedy–Thorndike experiment, the experiment of Ives and Stilwell on the 'transverse' (or more correctly the 'second order') Doppler effect [134; 135], and the previously-discussed observations of the lifetimes of cosmic ray μ mesons.

The Kennedy–Thorndike experiment could not be explained simply by the assumption of the Fitzgerald–Lorentz contraction of the apparatus in the direction of its motion through the ether, and in this sense it confirmed the relativistic nature of time. But as Ives [130] pointed out, both the Kennedy–Thorndike and the Michelson–Morley experiments *could* be accounted for without recourse to time-dilatation if somewhat different assumptions (from that of Fitzgerald and Lorentz) were made about the contraction of the interferometer arms.

The findings of Ives and Stilwell, of the Bell Telephone Laboratories in New York, (published in 1938 and 1941) gave a more direct confirmation of time-dilatation. According to classical theory, the optical Doppler effect occurs when there is motion of the light source toward or away from the observer concerned, but not when there is only transverse relative motion. Suppose, for example, that an observer O is at rest in the ether and that he observes a succession of

145

light flashes from a regularly emitting source. When the source is moving perpendicularly across the line of sight, successive signals will take equal times to reach O, and so the frequency of reception will be the same as the frequency of emission. The same consideration applies to the frequency of light waves from a moving source of continuous light; there is no observed change of frequency (colour) when the source moves transversely to the observer's line of sight. But the relativistic situation is different. The timekeeping of a source which moves at speed V relative to an inertial observer O appears slow by the factor $\sqrt{(1 - V^2)}$, i.e. approximately $1 - \frac{1}{2}V^2$, for small V. When the relative motion is transverse this is the only contributory factor to a change of apparent frequency. The observed colour-shift is towards the red end of the spectrum. It is called the *transverse* Doppler shift, and is a 'second order' effect in that (unlike the more familiar *radial* Doppler shift, due to mutual approach or recession of source and observer) it involves only the square of the speed V, and not V itself. Ordinarily, therefore, the transverse shift is very small and difficult to detect. Nevertheless, Einstein had suggested that it might be observable in the light emitted by the fast-moving particles in hydrogen canal rays.

The main difficulty in observing the transverse shift in such laboratory experiments is that of ensuring that the particles (produced in a discharge tube) are moving *exactly* transversely to the line of sight at the moment they emit the observed light. A very small deviation from this condition would render the results unreliable. Another difficulty is in obtaining a flow of particles whose speeds are all approximately the same, in order that the bands of the emitted light are not too diffuse for precise observation.

To overcome the first of these difficulties, Ives and Stilwell looked for a second order Doppler effect which was not really the transverse effect at all, though it was equivalent to it. They used the fact that the displacements in the wavelength of light when a source moves towards or away from an observer are not quite equal and opposite. For recession of the source the wavelength appears increased from, say, λ to $k\lambda$, while for approach of the source it appears decreased from λ to λ/k, where k has the same meaning as in §3 of Chapter 4, i.e.

$$k = \sqrt{\frac{1+V}{1-V}}. \qquad (5.1)$$

The average of the two displacements is therefore equivalent to a single displacement from λ to $\frac{1}{2}(k+1/k)\lambda$, where by (5.1)

$$\frac{1}{2}\left(k+\frac{1}{k}\right)\lambda = \frac{\lambda}{\sqrt{(1-V^2)}} \simeq \left(1+\frac{1}{2}V^2\right)\lambda, \qquad (5.2)$$

for V much smaller than 1.

The two experimenters decided to try to detect the small displacement $\frac{1}{2}V^2\lambda$, in (5.2) by observing simultaneously (using a mirror) the light emitted forwards and backwards from a beam of particles in a discharge tube, and comparing it with corresponding light from a source at rest. The problem of the diffuseness of the bands of emitted light was overcome with the development, in the 1930s, of discharge tubes capable of producing charged hydrogen particles travelling at nearly equal speeds. Results of one set of experiments were published in July, 1938, under the title 'An Experimental Study of the Rate of a Moving Clock', in the *Journal of the Optical Society of America* [134].

Note that a separate evaluation of V is necessary if the formula (5.2) is to be verified. This can be made by observing the magnitude of the ordinary first order Doppler shift in the light emitted in one direction, either backwards *or* forwards, or can alternatively be calculated theoretically from the voltage used in the discharge. The first column in Table 6 gives the voltages used in some experiments with hydrogen H_2. and the remaining three columns show respectively the values of $\frac{1}{2}V^2\lambda$:

(i) With V calculated from the voltage.
(ii) With V calculated from the observed first order Doppler shift.
(iii) Observed in the second order Doppler effect.

(The bottom row relates to a subsequent experiment; see below.) The wavelength λ is measured in Ångström units (1 Ångström unit = 10^{-10} m.)

The agreement of observational and computed values is seen to be extremely good. It should be noted that according to classical theory all entries in the last column would be zero, and so the experiment provides convincing confirmation of the time-dilatation phenomenon.

We recall that Ives frequently expressed a belief in the ether. It is thus only fair to point out that the two experimenters interpreted their results in terms of Lorentz's theory rather than Einstein's saying, 'The conclusion drawn from these experiments is that the change of frequency of a moving light source predicted by the Larmor–Lorentz theory is verified.'

TABLE 6

Some of Ives and Stilwell's observations on the second order Doppler effect in light from hydrogen H_2. (See text for explanation.)

Voltage	$\frac{1}{2}V^2\lambda$		
	Computed; (i)	Computed; (ii)	Observed; (iii)
7780	0·0203	0·0202	0·0185
9187	·0238	·0243	·0225
10574	·0275	·0280	·027
13560	·0352	·0360	·0345
18350	·0478	·0469	·047
42280		·1073	·1145

In the following year (1939), R. C. Jones [138] emphasized that the results were equally in keeping with Einstein's theory, and also drew attention to a possible source of error. Subsequent improvements in technique enabled Ives and Stilwell to use greatly increased voltages, leading to much larger shifts [135]. A typical result of the later experiments is shown in the last row of Table 6. It is interesting to note that while duly acknowledging Jones's technical contribution, the experimenters made no reference to the relativistic interpretation of their results, but reaffirmed their own conclusion.

Axioms by which the Michelson–Morley, Kennedy–Thorndike and Ives–Stilwell experiments lead to special relativity theory have been considered by a number of authors, including H. P. Robertson [292], A. Grünbaum [263], etc. The interpretations of the first two of these experiments, by Robertson, have already been given (pp. 34

and 35). For comparison purposes, Robertson's interpretation of the Ives–Stilwell experiments is stated:

> 'The frequency of a moving atomic source is altered by the factor $(1 - u^2/c^2)^{\frac{1}{2}}$, where u is the velocity of the source relative to the observer.'

Thus, whether we appeal to special relativity or not, the experiment provides direct evidence of the slowing down of one type of moving clock.

2. Mesons and the 'Fifties

Many experiments with laboratory-produced high speed particles became possible in the post-war years with the advent of large (and costly) accelerators. We shall consider here one or two actual or suggested experiments designed to measure the lifetimes of mesons and their dependence on speed.

At the Radiation Laboratory in Berkeley, California, E. Martinelli and W. K. H. Panofsky [173] measured the rest lifetime of the positively charged π meson, and published their results in 1950. They caused a cyclotron beam of protons to strike a carbon target in order to produce mesons; these sped from the target in all directions. Only mesons in a small range of directions were allowed through channels in shielding, and the experiment was concerned with these. The positively-charged mesons moved along spiral paths in the magnetic field of the cylotron, and the idea of the experiment was to determine their average lifetime by seeing how many were still intact after successive orbits had been traversed (Fig. 32). Actually, it was impracticable to make observations beyond one or two orbits, and this was probably the first time that sufficient energy had been available to permit observations beyond one orbit. But the more orbits that could be used, the more reliable the information gathered from the experiment.

The decay rate of a quantity of π mesons at rest in an inertial system is governed by the same form of law that applies to decaying particles generally, namely, that the rate of depletion at any moment is proportional to the number then remaining. Mathematically, this means that the number of particles still intact diminishes exponentially with the time. If the average lifetime of a meson is T, and if N_0 is the number which exist at the time $t = 0$, then the number N which remain at any later time t is given by the formula

$$N = N_0\, e^{-t/T}. \tag{5.3}$$

Martinelli and Panofsky were primarily concerned with the evaluation of T. In their rather early experiment of its kind relativistic effects were quite small, and T in (5.3) could be regarded almost

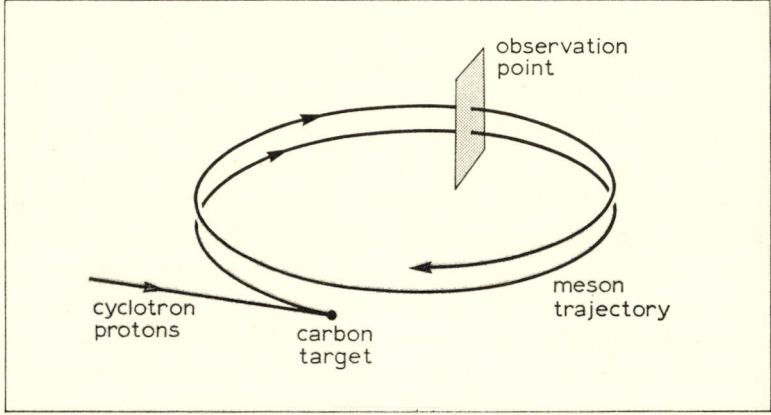

Fig. 32. *The fractional depletion of mesons during an orbit is used in lifetime calculations*

equally as the lifetime of the mesons at rest or at their orbital speed. They measured the proportion surviving each orbit by placing photographic plates in the way of the particles and observing the tracks made in the emulsion. By (5.1), if t_1 is the known laboratory time for the description of one orbit, then the fraction remaining is $e^{-t_1/T}$. This enables T to be calculated. In fact, their determination gave what is now regarded as too low a figure ($T = 1{\cdot}97 \times 10^{-8}$ s), and a more reliable value for the rest lifetime, $2{\cdot}54 \times 10^{-8}$ s, was obtained by another Berkeley group [271] using a different technique in 1957. But Martinelli and Panofsky did make the important suggestion that an accurate experiment with higher energy mesons might be used to test the relativistic ageing hypothesis.

If the energy of the mesons is sufficient for relativistic effects to be of consequence, and if the clock hypothesis applies to meson time-keeping, then in any one orbit only the fraction $e^{-t_1'/T}$ of those at the start will survive to the end, where t_1' is the proper time for the orbit as measured by a meson's 'clock'. In terms of the laboratory time t_1, we have $t_1' = t_1/\gamma$, where $\gamma = 1/\sqrt{(1-V^2)}$. Therefore,

$$\text{fraction surviving each orbit} = e^{-t_1/\gamma T}. \tag{5.4}$$

When γ is appreciably greater than 1, the fact should be reflected both qualitatively and quantitatively in the observations.

The relativistic effect was clearly evident in an equivalent, though somewhat different, high energy experiment performed by Durbin, Loar and Havens [80] of Columbia University, New York, in 1952, using the Nevis cyclotron. These workers found results which were in close agreement with other determinations of the rest lifetime of the positively charged π meson only if account is taken of time-dilatation. The importance of this aspect of the experiment of Durbin *et al* was stressed, in 1957, by W. Cochran [34; 35], who remarked that the experiment would not, however, overcome an objection made earlier by Dingle in respect of the cosmic ray μ meson observations [58]. In both cases, one-way journeys were involved and Dingle had insisted that time measurements over one-way journeys could not be used in evidence by the 'asymmetrical' school in the clock paradox controversy, because these measurements required the use of (conventionally) synchronized clocks.

Cochran therefore suggested a closed-path type of experiment with π mesons made to move along the required path by the presence of a magnet. (The arrangement would have much in common with that of Martinelli and Panofsky.) For comparison, the experimental arrangements of Durbin *et al* and of Cochran are shown in Figs. 33(a) and 33(b), respectively.

The magnet in the former case is used to deflect mesons of different energies through different angles, so that only those of approximately the same (known) energy follow a chosen path. In both experiments, C_1, C_2 and C_3 are 'coincidence counters' which record only the passage of mesons that go through all three (and hence do not register extraneous unwanted particles), and C_4 is a fourth coincidence counter which determines the proportion of surviving particles at the end of the observed path.

In suggesting a further variation, Cochran ingeniously argued that the result of the experiment would be a foregone conclusion, in view of the findings of Durbin *et al*! He considered a situation in

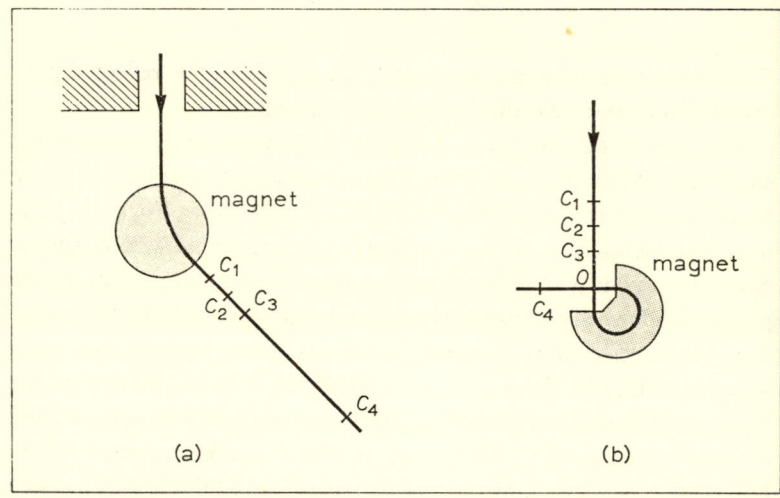

(a)

(b)

Fig. 33(a). *(after Cochran). The arrangement of the experiment of Durbin* et al

(b). *The arrangement of Cochran's first proposed experiment*

which the curved path in Fig. 33(b) is replaced by a rectangular path, with rounded corners. If the path length is L and the meson speed is V, then the proportion of mesons which survive the (laboratory) duration L/V of the journey is either:

$$p = e^{-L/VT} \quad \text{(if there is } no \text{ time-dilatation)} \quad (5.5)$$
$$p = e^{-L/\gamma VT} \quad \text{(if there } is \text{ time-dilatation)} \quad (5.6)$$

Note that (5.5) is also the proportion of an imaginary collection of 'stay-at-home' mesons, at rest at O, which would survive while the real ones make the round trip. Cochran argued that if the rectangular path were extended by an amount L' (by increasing the length of one pair of parallel sides) then the 'symmetrical ageing' school believing in (5.5) must expect a *further* reduction in the numbers surviving by the factor $e^{-L'/VT}$. On the other hand, the 'asymmetrical ageing' school believing in (5.6) expect their survivors to be further reduced by the different factor $e^{-L'/\Gamma VT}$. Since the findings of Durbin *et al* for straight line paths agree only with the latter reduction factor, the 'asymmetry' case is supported.

To complete this section we might usefully refer to some earlier experimental work, on cosmic ray μ mesons, which is particularly relevant to the clock hypothesis. Harold Ticho [224] described in 1947 some observations carried out at Climax, Colarado, at a height of 11 500 feet (3506 m) above sea level, to determine how the mean lifetime of the positively charged μ meson depended on altitude. His procedure was to make a direct measurement of the lifetime of individual mesons stopped in an 0·1 m aluminium absorber using a quartz crystal clock. A meson decaying inside an absorber emits an electron which almost always leaves the absorber, thus allowing detection of the decay from outside. Ticho found a rather low value, $1·78 \times 10^{-6}$s, but this was later explained in terms of an unexpected contribution from the disintegration of negatively charged mesons. When allowance is made for this contribution the expected value prevails, and so it appears that the enormously large accelerations during the braking in the absorber have no direct effect on ageing. Other workers [287] have also found that the average lifetime of the μ meson is unaffected by stoppage in various absorbers.

Crawford [41] in *Nature* in 1957, refers to a private communication from Ticho regarding the latter's experiment and a similar one in Chicago, at a height of 600 feet (183 m). He writes,

'If there were no asymmetrical ageing, Ticho would have observed at Chicago a rate anomalously reduced by a factor of about 40. Instead, Ticho observed roughly the expected number of decays at rest, both at low altitude and at high altitude. Thus, assumption 2 [that acceleration has no influence on the rate of 'ideal' clocks; taken by Crawford to include mesons] has been verified.'

153

3. The Mössbauer effect and the 'Sixties

Rudolf L. Mössbauer's discovery (made known in 1958) that an atomic nucleus can sometimes emit electromagnetic radiation, in the form of a γ ray, with effectively no recoil [286] paved the way for some of the most exciting techniques in modern experimental physics. Put at its simplest, the discovery showed how to produce radiation whose frequency is defined with almost incredible precision. A corresponding effect for the absorption of γ rays provides also a means of checking the same frequency to see, for example, whether it has undergone even a minute change between emission and absorption. The result is so important in experimental tests, in relativity and other branches of physics, that we shall start this section with an outline of the Mössbauer effect itself.

During the emission of ordinary light (i.e. a photon) by an atom which is initially at rest in a particular reference system, the atom gives up a definite amount of energy by falling from its initial 'state', A, to another 'state', B, of lower energy. The possible states of an atom have well-defined energies. Some of the energy released is spent in the recoil of the atom, just as when a gun is fired some of the explosive energy is spent in the recoil of the gun. Thus, the energy contained in the emitted photon is rather less than that necessary to allow its absorption if it encounters a second similar atom at the lower state B; there is insufficient energy to raise this atom to the state A. (In fact, the photon would really require a little *extra* energy, because there will be another recoil on its absorption, but a substance *can* absorb its own characteristic radiation if the temperature is right because the atoms have extra, and differing, energies due to thermal motion.)

A rather similar situation occurs with the emission of γ rays from a radioactive nucleus. But, for nuclei in crystalline solids, the crystal lattice may prohibit the separate motion of the emitting nucleus, if the energy available for recoil is too small. In this case, the recoil momentum is taken up by the crystal as a whole, and effectively all the released energy goes into the γ ray. (In the above analogue of the firing of a gun, this corresponds to the case where the gun is rigidly mounted on firm ground.) For example, French [262] quotes figures for an emission in a crystal of radioactive iridium ^{191}Ir, as used by Mössbauer in his original experiments. French calculates that for a crystal containing 10^{10} atoms (a very small quantity of the substance, weighing only 3×10^{-15} kg) the fraction of energy spent in recoil is only about 3×10^{-17}. Since the energy of a γ ray is known (theoretically and experimentally) to be directly proportional to its frequency, its frequency is confined within extremely close limits.

Of course, if the energy associated with an emission or absorption were *absolutely* fixed, then even such a minute wastage would be significant, and would prevent any subsequent absorption. But the energy is not completely fixed; there is simply a high probability that it will not deviate from a fixed mean value by more than a very small amount. In the above case of iridium this amount of deviation is of the order of 1 part in 10^{10}. Small though this is, the effect of recoil is negligible by comparison.

An absorber's response to a variation in the frequency of impinging γ rays may be examined by use of the Doppler effect. If a ray is emitted with frequency v, then according to an absorber which approaches the source at a small speed, V, the frequency appears increased to kv, where as usual

$$k = \sqrt{\frac{1+V}{1-V}} \backsimeq 1+V. \tag{5.7}$$

The greatest likelihood of absorption occurs when $V = 0$, but for other small values of V there is still an appreciable chance of this happening unless the magnitude of V exceeds a certain value. From what we have said above, this value is about 10^{-10}.

Only the relative velocity of source and absorber is important here. Mössbauer placed his iridium source at a point on the rim of a turn-table, which he rotated slowly (Fig. 34). The turn-table was surrounded by a lead shield outside of which was placed the absorber, and beyond this, a detector. A gap in the shielding allowed the passage of γ rays from the iridium crystal, S, only when the latter was directly approaching (or moving away from) the absorber, A. Those not stopped at A were detected at D.

Observations with different speeds of rotation showed that about

1 per cent of all γ rays from C were emitted without recoil, and that only half of these were stopped at A when the relative speed of C and A reached 0·02 m/s (i.e. roughly the speed of the tip of a second hand on a wall clock). This is equivalent to the value $V = 0·02/(3 \times 10^8) = 7 \times 10^{-11}$ in (5.7), and is in agreement with comments following (5.7).

Many later experiments based on the Mössbauer effect have been

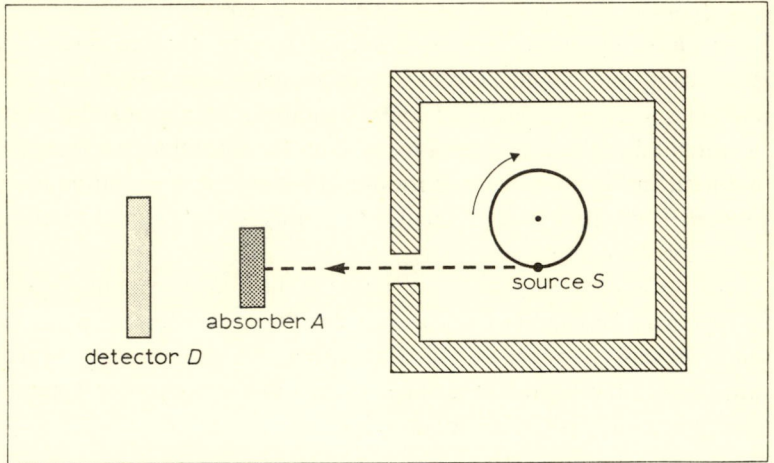

Fig. 34. *Mössbauer's experiment. When the rotating turntable moves the emitting crystal at a speed of 0·02 metres per second, the rate of absorption is found to be halved*

performed with other substances, such as the isotope of iron ^{57}Fe, in which 63 per cent of γ ray emissions are recoilless. Pound and Rebka, of Harvard, used this in a famous experiment [191] in 1960 relating to the principle of equivalence (§1, Chapter 6). They measured the change in frequency of γ rays as they 'fell' under gravity (the gravitational red shift) the absorber being at a different gravitational potential from that at the source. To restore resonant absorption by use of the Doppler effect in this situation, the absorber needs to be moved rather more slowly than the tip of the *minute* hand of the above-mentioned wall clock!

There are two ways in which the Mössbauer effect can be used in the experimental confirmation of time-dilatation; through its temperature dependence, and in connection with the transverse Doppler effect. The temperature dependence was pointed out by Pound and Rebka [190] and independently by a Cambridge undergraduate B. D. Josephson [139], in 1960. When the emitting or absorbing nuclei are vibrating thermally, a first order Doppler effect naturally occurs. But the to-and-fro motion of each nucleus (at speeds of the same order as those of jet aircraft) tend to cancel. According to Pound and Rebka,

> 'Thermally excited vibrations cause little broadening [of frequency bands] through first order Doppler effect under the conditions obtaining in the solid because the value of any component of the nuclear velocity averages very nearly to zero over the nuclear lifetime.'

The second order Doppler effect, which plays a similar part to that in the Ives–Stilwell experiment, is important. We know that if an emitting nucleus has speed v in the laboratory its timekeeping should appear slowed by the factor $\sqrt{(1-v^2)}$, or approximately $1-\frac{1}{2}v^2$ for small v, assuming time-dilatation. Now, according to classical kinetic theory the average value of $\frac{1}{2}v^2$ over all the nuclei in the lattice is proportional to the temperature T (°K) in accordance with the formula:

$$\tfrac{1}{2}v^2 = \frac{3kT}{2M} = 2{\cdot}4 \times 10^{-15}T, \qquad (5.8)$$

where M is the nuclear mass and k is *Boltzmann's constant*. Therefore, the frequency of the emitted γ rays appears reduced owing to the thermal motions by an average fraction $2{\cdot}4 \times 10^{-15}$ for each degree of temperature. Pound and Rebka point out that the classical formula (5·8) is not strictly accurate and that a closer figure on the right would be rather lower, about $2{\cdot}21 \times 10^{-15}T$.

If both source and absorber are at the same temperature, each will be affected in the same way by thermal motions, and therefore what is really important is a *temperature difference* between source and absorber. If the former is, say T_1°K hotter than the latter, then the γ radiation frequency will be below the optimum for absorption, by the fraction $2 \cdot 21 \times 10^{-15} T_1$. Pound and Rebka verified this by moving their absorber slowly towards the source so that first order Doppler effect compensated for the frequency reduction (Fig. 35). Their experimentally determined value for the fractional frequency reduction was $(2 \cdot 09 \pm 0 \cdot 24) \times 10^{-15}$ per degree, agreeing well with prediction.

Sherwin [211] drew attention in 1960 to the importance of this experiment, and one performed by a Harwell group, Hay, Schiffer,

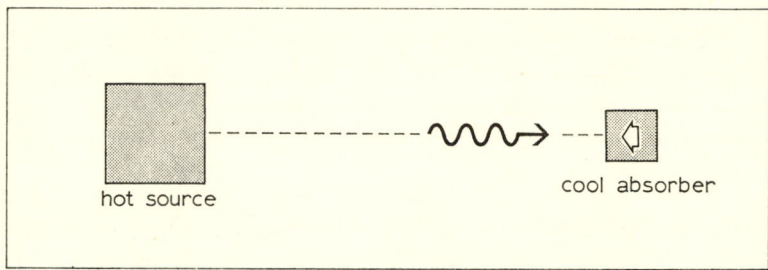

hot source

cool absorber

Fig. 35. *The stationary absorber finds the γ rays from the hot source too low in frequency for its taste. But the γ rays are palatable if it approaches the source at a suitable rate*

Cranshaw and Egelstaff [115] (see below), as evidence not only of time-dilatation but also of the clock hypothesis. In his opinion they provided 'the first experimental evidence of the timekeeping properties of accelerated clocks such as occur in the classic "clock paradox" of relativity'. The thermal vibrations of the lattice impart randomly oriented accelerations of the order of $10^{16}g$, to both source and absorber nuclei, and these accelerations themselves appear to produce no intrinsic frequency shift in ^{57}Fe to an accuracy exceeding 1 part in 10^{13}.

If there is any weakness at all in the argument of Pound and Rebka (who after all were not concerned with the clock paradox controversy), it would seem to lie in the assertion that the first order Doppler effects adequately cancel in the thermal motions of the nuclei. A very slight incompleteness in the cancelling would render the conclusion invalid. On the other hand, it would be very surprising indeed if the discrepancy turned out to be exactly of the right amount to give the false impression of an expected second order effect. (The cancelling question is discussed further by Sherwin.)

The experiment of Hay *et al* (also reported in 1960) was a direct test of the transverse Doppler effect. These workers used a circular band of cobalt ^{57}Co as a source, wrapped around a shaft between a pair of discs, while a larger diameter absorbing band of iron ^{57}Fe was placed concentrically near the edges of the discs (Fig. 36). The shaft was then rotated at up to 500 revolutions per second. Rotation

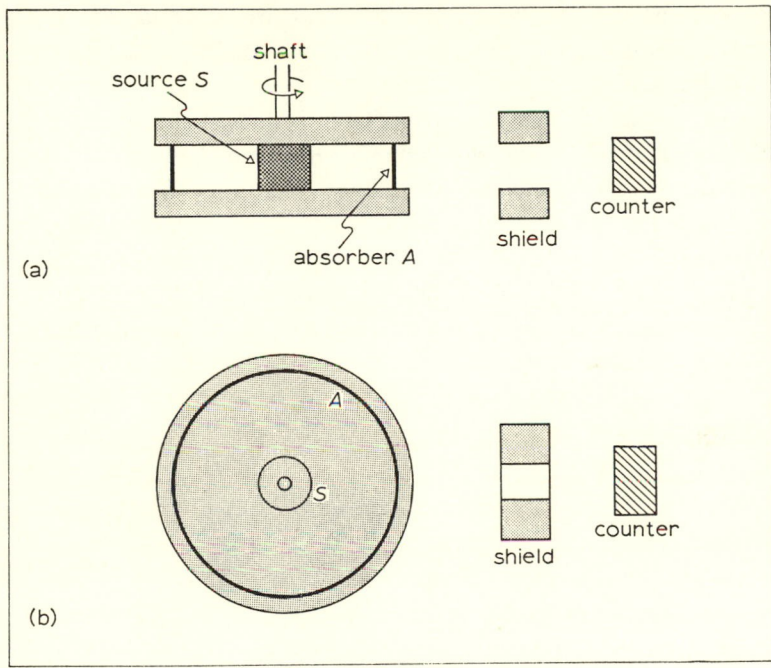

Fig. 36(a). *Schematic arrangement of the transverse Doppler effect experiment of Hay* et al
(b). *Plan view*

reduces the characteristic frequency of both the absorber (A) and the source (S), but by unequal amounts since the latter moves more slowly. If A and S have radii R and R_1, respectively, then their nuclei move at average speeds (expressed, as usual, as a fraction of the speed of light) of $R\omega/c$ and $R_1\omega/c$, where ω is the angular speed of rotation (radians per second). Thus, the γ rays from S reach A with frequency below the optimum for absorption by a fractional amount found to be

$$\tfrac{1}{2}(R^2 - R_1{}^2)\omega^2/c^2 = 2\cdot44 \times 10^{-20}\omega^2, \tag{5.9}$$

when values appropriate to the actual experiment are inserted. A rotational speed of 500 revolutions per second was expected to reduce the rate of absorption by 4 per cent, and Hay *et al* confirmed this experimentally by means of a counter which detected unabsorbed γ rays. One may regard this test as a further justification of the clock hypothesis, since fairly large accelerations are involved. The result can also be interpreted in terms of the principle of equivalence, as we shall see (Chapter 6).

One feature of this early experiment is that it did not test the *direction* of the frequency shift, although there is no known theory that would explain a time-*contraction* rather than a time-*dilatation* effect.

A somewhat different test was performed in 1961 by Champeney and Moon [32] with a ^{57}Co source and a ^{57}Fe absorber fixed at opposite ends of a high-speed rotor (Fig. 37). In this case, the timekeeping of each is affected equally by the rotation, and so the rate of detection

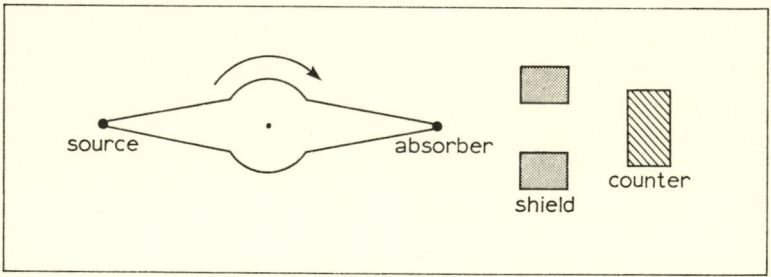

Fig. 37. *The experiment of Champeney and Moon*

at the counter should be independent of the speed of rotation. A main object of the experiment, performed at Birmingham University, was in fact to test the rotor system for vibration (which would show up in absorption rates through the Doppler effect). But a second object was to verify that *relative* transverse motion of source and absorber caused no second order Doppler shift. As expected, it did not. (The viewpoint of an observer moving with either accelerating end of the rotor needs a more subtle analysis!)

Further complementary experiments have since been made with either source or absorber at the centre of a rotor, and the other at one end [144; 145; 30]. Some of these experiments could distinguish also the direction of frequency shifts. (And some, in an incidental way, even confirm the absence of an ether wind past the earth down to speeds as low as 2 or 3 metres per second.)

Conventional relativists have concurred with the experimenters' interpretations of their results; but, not unexpectedly, others have dissented. Notable, perhaps, is the viewpoint of Essen [97] regarding the work of Champeney and Moon, and of Champeney, Isaak and Khan [31], in relation to Einstein's prediction of the slowing of moving clocks: 'As the authors point out, it is tacitly assumed that the acceleration has no effect. Einstein, however, goes on to explain that the effect, incorrectly predicted, is due to acceleration.' Without further comment we end this section with Essen's conclusion:

'When, as in the experiment of Champeney and Moon, the source and absorber have the same acceleration and are in relative motion there is no frequency shift. The most satisfactory inference is that the shift is due to the differences in acceleration, and not to the relative velocities. To bring the result into accord with the special theory the authors, it seems to me, make one assumption which contradicts the special theory and a second assumption which contradicts the general theory and the results of their own experiments.'

4. *The use of artificial earth satellites*

The use of artificial earth satellites for testing aspects of relativity theory has often been proposed. At the Tenth Annual Meeting of the Upper Atmosphere Rocket Research Panel at Ann Arbor, Michigan, in 1956 (the year before the launching of Sputnik I), S. F. Singer [212] reviewed some of the possible experiments. One would be a test of general relativity by means of the 'precession' of an elliptic satellite orbit (in which the elliptical path as a whole rotates). Such a precession has been observed, at near the limit of experimental accuracy, in the planet Mercury's orbit around the sun. Another would involve the effect of the gravitational potential on the rate of a clock in accordance with the principle of equivalence. Again, there is the effect which primarily concerns us here, the slowing of a satellite clock as observed from earth, due to the satellite's orbital motion.

According to the principle of equivalence, if a clock were held *at rest* above the earth in the vicinity of a satellite orbit, it should appear to run *fast* relative to a clock on the ground through its being at a higher gravitational potential (Chapter 6). Thus, a clock actually carried in a satellite is subject to two opposing influences on its rate, and these partly or wholly cancel. In very low orbits, the purely gravitational effect is minimal, whereas in an orbit at more than a certain height (for a circular one, the height is 3200 km above the ground) it predominates over the ordinary time-dilatation. Therefore, to measure the special relativistic effect alone low satellite orbits would be used, while high orbits would be needed to test general relativity or the principle of equivalence.

Singer suggested that satellite clocks should be compared with ground clocks by means of a 'counting' method in which a signal is sent to the ground only after a large number of 'ticks', because of unknown factors which could affect the transmission times of signals. A pair of caesium or similar clocks would be suitable for the experiment. Other methods, equivalent to a short-term comparison of *frequencies* by visual or radio observations are less reliable. Singer later pointed out that a low-altitude satellite experiment would involve the measurement of an effect of the order of 1 part in 3×10^9 (i.e. 1 second per century), but that atomic beam clocks could cope with this.

Banesh Hoffmann [119], at New York, examined the feasibility of the general relativity tests, and especially the extraneous effects due to the rotation of the earth and the lack of symmetry of its gravitational field. He believed that the requisite measurements (relevant here because they need the same order of precision as the measurement of the special relativistic time-dilatation) would be possible. M. Subotowicz [217] held a similar view, saying that observations need only be extended over about one year.

The use of *masers* as clocks, governed by properties of ammonia molecules, was considered in 1957 by Møller [181] who noted that they were not sufficiently accurate for terrestrial tests of general relativity using mountain heights of 3 km. But he remarked,

'Although it sounds somewhat fantastic at the moment, it may well be that, before clocks of considerably higher accuracy are constructed, it will be possible to use artificial satellites, in which case the effect in question can be made a thousand times bigger.'

At the time of writing this book, only a dozen or so years after the publication of Møller's investigation, the tests are certainly possible. But there have also been, in the meantime, enormous advances in other areas of science and technology. The 'Mössbauer revolution' has enabled simpler tests of relativistic time phenomena to be made, and consequently the attraction of using satellites for this purpose is much reduced. (For further discussion of satellite tests, see the next section.)

Chapter 6

THE CLOCK PARADOX IN GENERAL RELATIVITY

'A good many physicists believe that this paradox can only
be resolved by the general theory of relativity. They find great
comfort in this, because they don't know any general relativity'.

PROFESSOR ALFRED SCHILD, *The American Journal of Physics.*

1. The principle of equivalence

In a public lecture, a few years ago, Herman Bondi observed, 'When a physicist is out birdwatching and falls from a cliff, he never worries about his binoculars; he knows they will fall alongside him'. Bondi was illustrating, in a picturesque way, the well-known *law of Galileo,* that all bodies fall with equal acceleration in a gravitational field ('all bodies are falling with equal speed'), provided that non-gravitational forces such as air resistance are negligible. Galileo's experiments, performed with wood and lead weights (rather than physicists and binoculars) dropped from a high tower at Pisa, began towards the end of the sixteenth century, although as R. H. Dicke [255] has pointed out, the law was known earlier, and primitive man probably noticed that a lizard knocked out of a tree, by a stone, fell to the ground almost simultaneously with the stone.

In Newtonian mechanics, the law of Galileo is expressed by the assumption that the gravitational mass of a body is proportional to its inertial mass. The gravitational mass, m_G, is defined in terms of its attraction to other massive bodies, and the gravitational masses of different bodies may be compared by weighing them. The inertial mass, m_I, is essentially quite different, it being a measure of the resistance of the body to acceleration by the application of a given force (such as a certain muscular effort, or the tension in a particular spring when stretched to a specified length). Different bodies have equal accelerations under gravity only if the gravitational force,

164

determined by m_G, is proportional to the resistance to acceleration, determined by m_I. One unsatisfactory feature of Newtonian gravitational theory is that while the strict proportionality of these two kinds of mass is accepted, it is in no way an essential or integral part of the theory (as it is in general relativity).

Galileo's law is essential to general relativity theory, and it is worth considering how well the law is supported by more modern precise experiments. The best known of these were performed by the Hungarian, Baron Roland von Eötvös, first in 1889 and again in later years [258, 259]. Eötvös's procedure may be illustrated by considering the small deviation from the vertical of a pendulum (plumb line) hanging at rest in the laboratory. The angle at which the pendulum hangs is determined by two forces acting on the bob, the gravitational attraction towards the centre of the earth and the centrifugal force directed away from the earth's axis of rotation. While the former force is gravitational, the latter is inertial, since it represents the tendency of the bob to move in a straight line tangential to the earth's surface. Thus, if two pendulums had dissimilar bobs for which the ratios m_G/m_I were unequal they would hang at slightly different inclinations to the vertical.

Eötvös attached weights of different materials to a light horizontal balance arm which was itself suspended by a thin platinum-iridium wire. Except when the balance arm lies in the earth's meridian through the laboratory, any differences in the ratio m_G/m_I for the two weights would appear as a tendency to twist the suspension wire through a small angle. This could be detected by rotating the apparatus through 180°, so that the roles of the two weights is interchanged and the direction of twist reversed. Eötvös detected no such twist of the suspension wire relative to the rest of the apparatus, and concluded that the acceleration of his weights under gravity would be equal to within the accuracy of his determination. It is a feature of torsion balances that extremely precise measurements can be made with them, and Eötvös himself claimed an accuracy of about 5 parts in 10^9. (Dicke has suggested that this might be a rather optimistic claim. For example, if the Baron came too close to the apparatus, his own mass could have affected the result by an amount much in excess of the claimed experimental error!) Nevertheless, there is no doubt that Eötvös's experiments, using weights made from a variety of materials such as brass, cork, glass, etc., very strongly supported Galileo's law.

Another result of interest is that of L. Southerns [299] who confirmed the law for uranium, thus demonstrating that the inertial mass equivalent of the nuclear binding energy also possesses an appropriate gravitational mass.

Dicke and his colleagues at Princeton pushed the accuracy of the the Eötvös experiment still higher in the early 1960s [255]. For various technical reasons they made use of the inertial forces arising from the acceleration of their apparatus towards the *sun*, rather than that from the earth's rotation. Their device consisted of a horizontal triangular balance with weights (two of aluminium and one of gold) at the vertices, suspended by a quartz fibre, all enclosed in an evacuated container. Great precautions were taken to eliminate spurious effects. The whole apparatus was buried in a 12 feet-deep pit and left undisturbed for long periods (the necessary rotation being carried out by the earth itself), while monitoring was automatic and continuous. Dicke's group concluded that 'aluminium and gold fall toward the sun with the same acceleration, the accelerations differing from one another by at most 1 part in 10^{11}'. This represented an accuracy of at least 200 times that of Eötvös.

Galileo's law is vital to the *principle of equivalence*, according to which a gravitational field is locally indistinguishable from an acceleration of the reference system. In 1911, Einstein [84] set forth this principle for the first time, arguing as follows. Consider two reference systems. The first, K, is at rest in a uniform gravitational field, as may be considered to exist over a small region of the earth's surface, the field being such as to cause free bodies to move with acceleration g in the negative z direction (Fig. 38a). The second system, K' applies in a region of space where there is no gravitational field, and is accelerated at a constant rate g in the direction of its z' axis (Fig. 38b). (Difficulties in defining accelerating reference systems are of no concern here; the Newtonian approximation will suffice.)

In the two systems, the motion of free bodies is exactly the same. In the one case they are considered to be accelerated under the action of a 'true' gravitational field; in the other, there is an 'apparent' gravitational field due to the acceleration of the reference system. Thus, there appears no difference between the 'true' and 'apparent' fields provided that Galileo's law is *strictly* obeyed, though even a small difference in the falling rates of different substances in gravitational fields would render the argument invalid.

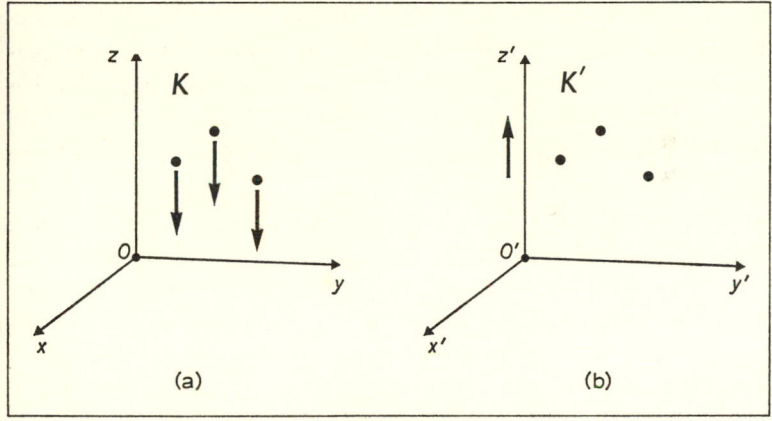

Fig. 38(a). *The reference system K is at rest in a uniform gravitational field. Free bodies fall with constant acceleration g .*

(b). *The reference system K' is accelerated at constant rate g in the absence of a gravitational field. Free bodies 'fall' with constant acceleration g*

However, the assertion that the systems K and K' are equivalent in every respect is much more than a restatement of Galileo's law, which tells us nothing about non-gravitational phenomena. It was not known in 1911 whether, for example, experiments in electromagnetism would reveal a genuine distinction between the two systems. Einstein suggested there was no valid distinction, saying:

'This experience, of the equal falling of all bodies in the gravitational field, is one of the most universal which the observation of nature has yielded; but in spite of that the law has not found any place in the foundation of our edifice of the physical universe.

'But we arrive at a very satisfactory interpretation of this law of experience, if we assume that the systems K and K' are physically exactly equivalent, that is, if we assume that we may just as well regard the system K as being in a space free from gravitational fields, if we then regard K as uniformly accelerated.'

Another way of expressing the equivalence is the statement that a reference system which is falling freely in a uniform gravitational field is indistinguishable from an inertial system. This could be confirmed by a strong-willed lift-attendant who makes careful observations between the time his lift cable breaks and the moment he crashes to a stop. But gravitational fields in nature are not uniform. The field near the surface of the earth is directed radially inwards towards the centre; and gravitational fields generally are irregular. Thus, the principle of equivalence is strictly local in character, and applies only in small regions in which the variation of strength and direction of the field is negligible. In an individual approach to general relativity, Fock [103] has emphasized the distinction between the law of the proportionality of inertial and gravitational masses, and the principle of equivalence. The former is an integral part of general relativity in that the theory specifies the paths of (small) falling bodies independently of their constitution. It thus is 'a fundamental law of general character whereas the principle of equivalence is strictly local'. (Fock further asserts, however, that the principle is 'not unrestrictedly part of Einstein's theory of gravitation, as expressed by the gravitational equations'; this is a rather extreme viewpoint.)

We shall now assume the principle to be valid and use it in an approximate calculation to determine the effect of a gravitational field on the rate of a clock. Suppose that a standard clock, A, is placed on the ground and that a second one, B, is at rest at a small height H vertically above A. If B emits n light signals during each unit of its time, at what rate are these received at A? (In this completely static situation, successive signals must be regarded as taking equal times to travel from one clock to the other, and hence the rate of reception of the signals provides A's measure of B's rate.)

In Fig. 39(a), we have chosen K to be the (approximately inertial) reference system of earth, with z axis vertically upwards. An 'equivalent' situation is one in which there is no gravitational field, and A and B are rigidly separated at distance H apart, both accelerating at rate g in the direction AB. Let K_0 be an inertial frame in which A and B are initially at rest, and in which there is no gravitational field. We need consider only the first few signals. In K_0, these signals from B will take very nearly a time H (at speed 1) to reach A, since A

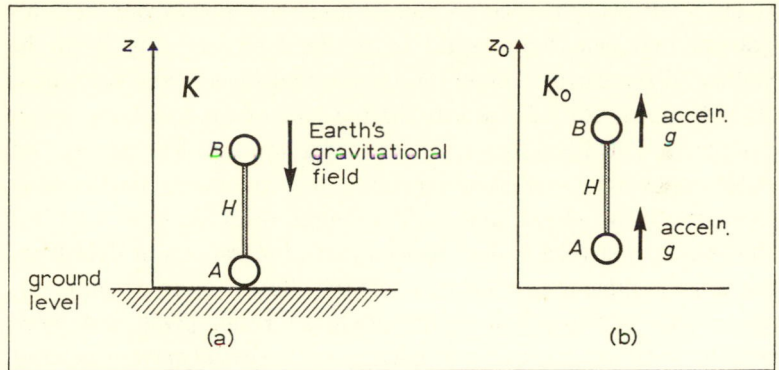

Fig. 39(a). *Clock B is at rest at height H above clock A, which is at ground level*

(b). *Clocks A and B are at a fixed distance H apart and are accelerating at rate g in the system K_0 in which there is no gravitational field*

moves little during the transit of a light-signal. Thus the signals arrive when A's speed is gH (in the Newtonian approximation). By the Doppler effect formula, the rate of reception is increased due to A's motion by a factor $1+gH$, and so his rate of reception is

$$n' = (1+gH)n. \qquad (6.1)$$

The quantity gH is the amount by which the gravitational potential at B exceeds that at A in the original problem (Fig. 39a). A more detailed analysis shows the general result that a standard clock always runs fast, by a factor $1+U$, when fixed at a nearby point where the gravitational potential is higher than at the point of observation by an amount U. Similarly, if a clock is at a point of lower potential than the observation point it appears to run slow, and the factor $1+U$ again applies but with U now negative. Thus, if we slowly raise a clock to the top of a building, and after a long time lower it back to the ground, it should be found to have gained during the process.

As an example of a clock mechanism we can consider the vibrations of an atom, and the frequency of its emitted light. The gravitational potential near the surface of the sun is lower than that of the earth (more work being needed to remove a massive body from the sun's gravitational field than would be regained on lowering it to the surface of the earth). Therefore, the spectral lines in the sun's light should appear shifted towards the red end of the spectrum, when compared with light from our terrestrial sources. The amount of reddening is very small, corresponding to a frequency shift of only two parts in a million, and early attempts to detect this solar red shift were unreliable. It then seemed that observations of light from the dense companion to Sirius, where the effect is more pronounced, gave better confirmation of the predicted result [295]. But more recently J. Brault, a student of Dicke, used a special spectrometer of his own design to make a new measurement of the solar red shift, and confirmed the predicted amount to within an accuracy of about 5 per cent [255].

The formula for the gravitational red shift can also be obtained by assuming the law of the conservation of energy. If a photon is emitted with frequency v_B from B, then its mass m and energy E are given by the relations

$$m = E = hv_B \tag{6.2}$$

where h is Planck's constant. As the photon 'falls' from B to a point of lower potential, it gains energy from the gravitational field. On arriving at A, the energy gained is mU, giving it a total energy of

$$hv_A = hv_B + mU \tag{6.3}$$

On replacing m by hv_B (by 6.2), we get the expected result:

$$v_A = v_B(1+U).$$

We note also that the rotor experiments using the Mössbauer effect, described in Chapter 5, can be interpreted in terms of the principle of equivalence. The centrifugal force on a particle of unit mass, fixed on a rotor arm at distance r from the centre, is $r\omega^2$, where ω is the angular speed of rotation. The work done by such a force if the particle is moved outwards from radius R_1 to radius R is

$$\omega^2 \int_{R_1}^{R} r \, dr = \tfrac{1}{2}\omega^2(R - R_1),$$

which can be regarded as the drop in potential, between R_1 and R, of the outward 'apparent' gravitational field in the rotating system of the rotor. The result (5.9) follows immediately when account is taken of the different time units used there.

Now let us turn to the question of the rates of satellite-borne clocks. Two main effects have to be considered, one gravitational and one due to Doppler shift. If we regard the earth as perfectly spherical, then according to Newtonian theory the gravitational potential at an exterior point at distance r from the centre is $-GM/r$, where M is the mass of the earth and G is the gravitational constant. The potential at radius r therefore exceeds that at the surface of the earth (radius r_0) by the amount

$$U = GM\left(\frac{1}{r_0} - \frac{1}{r}\right).$$

A clock held at radius r, at rest in a system following the earth (but not rotating), would be found from the ground to run fast by the factor $1 + U$, for this value of U.

For a satellite in a circular orbit, the speed in the non-rotating system is, according to Newtonian dynamics, $v = \sqrt{(GM/r)}$, which gives rise to a time-dilatation factor

$$\sqrt{(1 - v^2)} \simeq 1 - \tfrac{1}{2}v^2 = 1 - GM/2r,$$

acting in opposition to the gravitational effect. Ignoring the relatively slow motion of ground clocks, in the present reference system, we find that satellite clock rates are fast (or slow) by the factor

$$(1 + U)(1 - \tfrac{1}{2}v^2) \simeq 1 + U - \tfrac{1}{2}v^2 = 1 + GM\left(\frac{1}{r_0} - \frac{3}{2r}\right). \qquad (6.4)$$

If the height of the satellite happens to be just 3200 km (one-half the earth's radius), r is equal to $3r_0/2$, and the value of (6.4) is 1. In this critical case the two opposing effects cancel, and the satellite

171

clock keeps time with those on the ground. If above this height, the satellite clock gains (suggesting a means of violating McCrea's principle of impotence, though by a very small amount!) and if below the critical height it loses. In Fig. 40, the dependence of rate (seconds per century) on height (km) is shown.

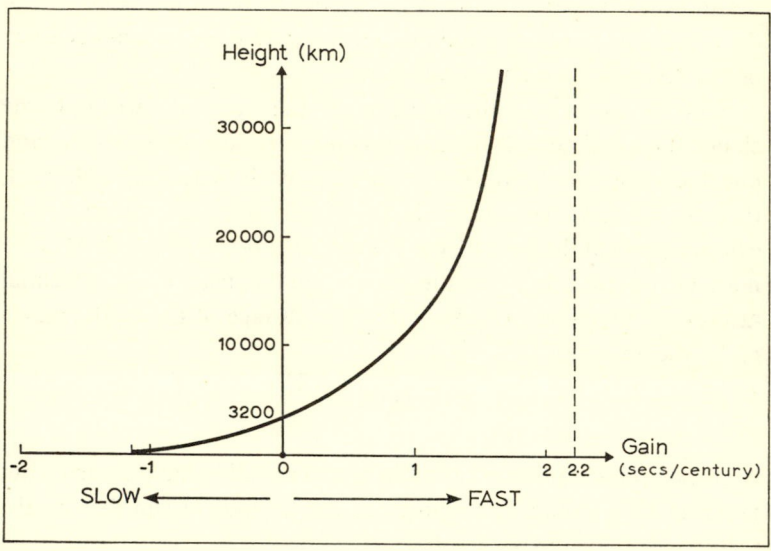

Fig. 40. *Dependence of the rate of a satellite clock on the height of its circular orbit*

Cochran [36] has pointed out that another instance of a 'moving' clock gaining with respect to terrestrial clocks occurs when the former is thrown vertically upwards and caught again like a ball. (Experimentation here would be very difficult.)

Banesh Hoffmann [120] has considered measuring the difference in rates of a *terrestrial* clock at noon and at midnight, due to the difference in its distances from the sun at these times. He finds that the effect is virtually cancelled by the Doppler shift, but he notes that if the principle of equivalence were not valid by the amount of possible error in the Pound and Rebka experiment (about 10 per cent), then his own experiment might detect this. Jointly with W. T. Sproull he considered also the change in rates as the earth-sun distance varies due to ellipticity of the earth's orbit [121]. This is a not inconsiderable change (the maximum fractional variation is about $4 \cdot 9 \times 10^{-16}$) but its measurement involves the awkward comparison of rates experienced by a single clock at six-month intervals.

We have thus far used the principle of equivalence to investigate the behaviour of clocks in gravitational fields. Later in this chapter we shall consider its use in treatments of the clock paradox.

2. A roundabout approach to general relativity

The study of rotating reference systems, such as that of a revolving roundabout, played an important part in the origins of general relativity theory. It will be useful to consider these systems briefly as an approach to the theory.

In special relativity, as in Newtonian theory, rotations are absolute, i.e. are relative only to a pre-determined, rather abstract collection of inertial systems. Einstein, however (like Bishop Berkeley, Ernst Mach and others) believed that the concept of the rotation of a body was meaningful only in respect of its motion relative to other bodies. But suppose that the centrifugal force felt by a roundabout rider is attributed to the rotation relative to distant matter (in the form of galaxies). Should not the laws of physics predict a corresponding outward force on passengers on a 'stationary' roundabout, about which the distant galaxies were caused to rotate? Einstein's first target in the construction of the new theory was to remove any distinction in these two aspects of the same relative motion: 'The laws of physics must be of such nature that they apply to systems of reference in any kind of motion' [85].

The notion of an accelerating system (including one in rotation) is, we know, hard to define because there is no really satisfactory definition of rigidity for the 'scaffolding' of the reference frame. The best way out of this difficulty is to impose as little restriction as possible on admissible systems, and to allow almost any imaginable scheme of labelling events in the formulation of physical laws. All such schemes are thus to be equally acceptable. Every event is to be marked by a set of four numbers, or *coordinates* (x^1, x^2, x^3, x^4), of which the first three have a certain space-like character and the fourth a time-like character. This, in effect, means that if the co-ordinates of two events are identical, except for a small difference in their x^4 values, then the events in some sense occur at different *times* rather than at different *places*. (More accurately, one supposes that an observer could, in principle, move so as to be present at both events, and that his clock would record different times for them, the later event being the one with larger x^4 coordinate). It is *not*, however, assumed that x^4 is the reading of any standard clock.) On the other hand, any two events with equal x^4 coordinates, but with slightly differing values of their x^1, x^2 or x^3 coordinates, occur simultaneously but at different *places* in the local inertial system of a certain freely-falling observer. It is in this sense that the first three coordinates are *space-like*.

The statement that any such coordinate system is as good as any other in the formulation of physical laws is the *principle of general covariance*, originally formulated in Einstein's (1916) paper ('The Foundation of the General Theory of Relativity') in the form [85]:

> 'The general laws of nature are to be expressed by equations which hold good for all systems of coordinates, that is, are co-variant with respect to any substitutions whatever.'

In fact, it is now realized that with sufficient ingenuity almost any theory (including Newtonian theory) can be expressed in generally covariant form.

General relativity, on this basis, arose through the combination of the two principles, of *general covariance* and of *equivalence*. According to the latter, the apparent gravitational field resulting from the acceleration of a reference system is to be locally indistinguishable from a true gravitational field. Therefore, any theory which can cope satisfactorily with arbitrary coordinate systems should also cope with gravitational fields in an integral way, as is not the case in the Newtonian and special relativity theories.

What the study of the rotating roundabout suggested, was that the new theory should be geometrical. We can see this from Einstein's following example. Imagine a circle drawn on the ground, adjacent to the circular rim of the roundabout, and suppose that a ground observer and a roundabout rider each attempt to determine the value of the constant π by measuring the circumference and radius of these respective circles with identical short measuring-rods. In the inertial system, S, of the ground let m rods laid end-to-end extend around the circumference of the circle on the ground, and let n rods end to end be needed along a radius (Fig. 41). Then,

$$\frac{m}{2n} = \pi = 3 \cdot 14159 \ldots .$$

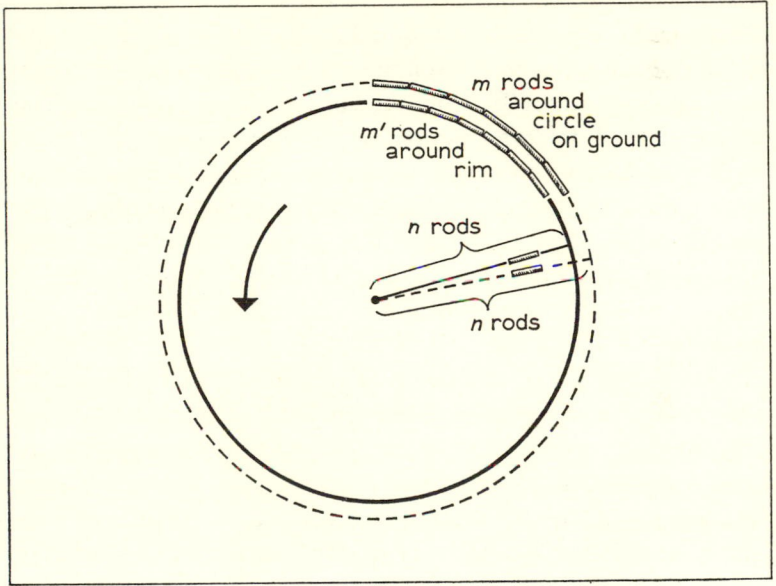

Fig. 41. *Solid lines relate to the roundabout and broken lines relate to the ground. The two circles are nearly coincident*

On the roundabout, let the number of rods needed around the rim be m'. Observed in S, each of these rods is moving along its length, parallel to one of the m rods on the ground, and therefore exhibits Fitzgerald–Lorentz contraction. More of the contracted rods are needed to extend around (effectively) the same circle, and so m' is greater than m. This, of course, is an absolute result, not dependent on who compares the two numbers. Now consider rods laid along radii in the two systems. In S, the roundabout rods move transversely to those on the ground, and there is no Fitzgerald–Lorentz contraction. (At least, this is the case if we assume that the length of a short rod depends only on its speed and not its acceleration – the *rod hypothesis*. Note that the terrestrial experiments which confirm the contraction formula were carried out in just such a rotating system, that of the earth, where comparable accelerations were present.) Therefore an equal number of rods extends along a radius in either system. It follows that the roundabout calculation for π would give the oversize value $m'/2n$.

The conclusion to be drawn from this imaginary experiment is that the spatial geometry on the rotating roundabout, as determined by standard measuring rods, is not the same as on the Euclidean plane. We should find, for example, that if the rider chooses three points A, B, C on the roundabout as vertices of a triangle, whose sides are to be formed by geodesics (lines of shortest length according to his measurements), then the angles at A, B and C do not add up to exactly 180°, but to rather less than this value. (Geodesic triangles on the earth's surface, formed by arcs of great circles, have the *opposite* property.)

A fuller discussion of rotating roundabouts is given by Arzeliès [2], who answers many possible objections, and in references that he cites. Perhaps worthy of note here is a peculiar timekeeping property. Let a large number of standard clocks, C_1, C_2, . . ., C_n, be carried on the roundabout, equally spaced around a circle concentric with the rim. If the radius of the circle is r, and the angular speed of rotation is ω, then each clock will be observed in S to be slowed by the dilatation factor $\sqrt{(1-r^2\omega^2)}$. But if the rider successively attempts to synchronize them using Einstein synchronization he is doomed to failure. Let him synchronize C_2 with C_1, then C_3 with C_2, and so on around the circle. When he has dealt with C_n, it will be found to be out of synchronization with the neighbouring clock C_1. If such rotational effects were more pronounced than they are, at low speeds, life on earth would be very complicated indeed!

An essential distinguishing feature of the Euclidean plane is the fact that Pythagoras's theorem is generally valid. In particular, if dx, dy denote the differences in the cartesian coordinates (x, y) of neighbouring points, then the distance dl between the points is given by

$$dl^2 = dx^2 + dy^2. \tag{6.5}$$

Only when the geometry of a surface is that of the Euclidean plane can its points be labelled (x, y) in such a way that the distance formula (6.5) is valid. Thus, this is not possible on the surface of a sphere, or on a rotating disc.

The geometry of the two-dimensional Minkowski space-time (x, t) is 'pseudo-Euclidean' and has some analogous features to that of the Euclidean plane. Here, the quantity corresponding to dl in (6.5) is not an ordinary distance, but a real or imaginary quantity ds, associated with neighbouring events whose coordinates differ by dx, dt, where

$$ds^2 = dt^2 - dx^2. \tag{6.6}$$

When the events in question are on the world-line of a moving body, dx/dt is the instantaneous speed, v, and by (6.6)

$$\begin{aligned} ds &= \sqrt{(dt^2 - dx^2)} \\ &= dt\sqrt{\{1 - (dx/dt)^2\}} \\ &= dt\sqrt{(1 - v^2)}, \end{aligned} \tag{6.7}$$

which is the proper time-interval between the events. The proper time does not depend on the particular inertial system employed in its calculation and so in a different system (x', t') we obtain the same value for the quantity ds from the formula

$$ds^2 = dt'^2 - dx'^2, \tag{6.8}$$

(as may also be verified from the Lorentz transformation relations (2.9)).

The basis of general relativity is that four-dimensional space-time is *Riemannian*, which means that its geometrical properties are described by means of a measurable quantity ds associated with any pair of neighbouring events, whose coordinates (of the general type described earlier) differ by amounts dx^1, dx^2, dx^3, dx^4, and that ds^2 is quadratic in the dx's. It is further assumed that (as in the case above) the value of ds is independent of the coordinate system in which it is calculated and is again the proper time-interval when the events are

177

on a world-line. In the total absence of gravitating matter the space-time is the pseudo-Euclidean space-time of Minkowski, for which the *metric ds^2* is an immediate extension of (6.6),

$$ds^2 = dt^2 - dx^2 - dy^2 - dz^2 \qquad (6.9)$$

in appropriate coordinates of any inertial system.

In the presence of gravitating matter the geometry is assumed to be distorted in a definite way, as prescribed by gravitational equations. To incorporate Galileo's law, it is necessary to suppose that the motion of free test bodies takes place along paths which are singled out by the space-time itself, since the paths are not to depend on the mass or composition of the test bodies. Few such classes of paths are singled out in a space-time. As a generalization from the inertial motion of test bodies in Minkowski space-time, it is natural to assume (as the general theory does assume) that the world-lines of free test bodies are always (time-like) *geodesics*. (If the proper time-interval ($\int ds$) along a particular world-line between events A and B, is greater than that along any other conceivable world-line joining the same events, then the former path is called a time-like geodesic.) The assumption of geodesic motion for freely falling bodies fits excellently with observation. However, the equations which determine these space-time paths may have other solutions – again called geodesics – which are not those of greatest proper time.

There is a familiar analogue in regard to non-flat two-dimensional surfaces. Here, shortest paths are geodesics. The shortest route from, say, London to New York may be determined by stretching a thread between the positions of the cities on a globe. It will, of course lie along a great circle arc, and all shortest routes between pairs of points on a globe are along arcs of great circles. But a second great circle arc from London to New York also exists (along the remaining greater part of the original great circle) and does not represent a shortest route. The same mathematical equations throw up the two types of path without distinction.

Were it not possible for such cases (of time-like geodesics which are not paths of greatest proper time) to exist in general relativity, the theory would certainly predict the maximum ageing between two events for an observer in free fall between them. In fact, this is always the case if the events are near together and we compare the free-fall path with only *slightly* different ones, as might be followed by spaceships with *very* weak motors. An analysis of the conditions for the geodesic paths to be one of greatest proper time has been given by the late Robert Boyer [14].

Mikhail [176] has considered an example of two clocks that separate and reunite though each is freely falling in the interim. The situation envisaged is like that examined by Cochran (§1). A spaceman in a circular orbit around the earth throws a clock upwards, i.e. radially away from the earth, and catches it next time round. Here, the world-lines of the satellite and the thrown clock are both geodesics in the exterior space-time of a spherical gravitating body, for which the metric (due to Schwarzschild) is

$$ds^2 = \left(1 - \frac{2m}{r}\right) dt^2 - dr^2 \bigg/ \left(1 - \frac{2m}{r}\right) - r^2\, d\theta^2 - r^2 \sin^2 \theta\, d\phi^2$$

where m is a mass constant, r is a radial coordinate, θ and ϕ are latitude and longitude, and t is a time coordinate. Mikhail finds, as Cochran did, that a satellite clock registers less time for the duration of separation than the thrown clock.

In our treatment of the twins paradox in special relativity we placed great emphasis on the dynamical asymmetry due to the fact that one twin and not the other used rocket motors. But now we find, in the presence of a real gravitational field, that asymmetrical time passage can occur even when neither twin uses motors. Asymmetry is present in both the former and present problems, however, in the aspect presented by the universe to observers following different space-time routes.

3. Paradox revisited

We shall consider finally some attempts that have been made to 'resolve' the clock paradox in Minkowski space-time by applying general relativity techniques. The notation is that of §1, Chapter 4.

The first discussion of the problem in terms of the general theory appears to have been given by Einstein in 1918 [86]. He produced no detailed calculations, but described in general terms the viewpoint of the traveller M. In any period of acceleration, due to the firing of rockets, M can regard himself as held at rest by his motors against a uniform gravitational field, according to the principle of equivalence. Since R uses no motors, he falls freely in this gravitational field. Let us consider just the period of turn-round at the half-way stage: R is accelerated towards M by the field and so is at a point of higher gravitational potential than M. But clocks at higher gravitational potentials run faster, and so the time recorded along $M_1 N_1$ exceeds that along $M' N'$ in Fig. 24. The paradox no longer exists.

A detailed, though approximate calculation of the timekeeping during the various stages of the journey in M's experience was given by Tolman, in 1934, on the above lines [227]. However, the argument is circuitous, as Builder [18] has strongly emphasized. For, the dependence of the rate of a clock on gravitational potential was found by Tolman by considering observations in an accelerated system, using the principle of equivalence as in §1. If the formula so obtained is applied to the apparent gravitational field due to M's acceleration, nothing can be deduced that could not be found directly by inspection of M's system without the introduction of the apparent field. According to Builder:

> 'This tortuous procedure succeeded in hiding the paradox rather than resolving it, for it scarcely need be pointed out that the procedure would be quite invalid if the restricted theory were indeed not properly applicable to the problem considered.'

In 1943, and in a similar discussion later in his book, Møller [178; 179] gave a complete and exact treatment by introducing a rigid accelerating reference system, like the one described in §1, Chapter 4. His procedure can be illustrated by reference to the situation in Fig. 25. The substitution (see 4.4)

$$1+ax = (1+aX)\cosh aT, \qquad (6.10)$$
$$at = (1+aX)\sinh aT \qquad (6.11)$$

brings the Minkowski two-dimensional space-time metric (6.6) into the form which applies in the X, T coordinates of the accelerated system, i.e.

$$ds^2 = (1+aX)^2\, dT^2 - dX^2. \qquad (6.12)$$

(The substitution is not, in fact, permissible for the whole of the Minkowski space-time, but it is valid throughout the region of interest.) The world-lines for which X is constant are those of the hyperbolic motions depicted in the figure.

A comparison between the rates of clocks 'at rest' at different points of the accelerated system can be made directly from (6.12). On any such world-line, $X = X_1$, the proper time-interval ds is related to the coordinate time-interval dT by

$$ds = (1 + aX_1)\, dT.$$

On M's world-line in Fig. 25, $X = 0$, and so

$$ds = dT.$$

Thus, by comparing these equations, we see that in M's system the clock at $X = X_1$ runs faster than his own by the factor $1 + aX_1$.

Møller examined the motion of free test bodies in the X, T system and defined a gravitational potential by comparison with Newtonian theory. (The possible motions can be determined directly from (6.10) and (6.11), since they are known to be straight lines in the xt space-time diagram, without recourse to the geodesic equations for the metric (6.12).) The potential turned out to be approximately aX, for small values of X, as the above result indicates. Having already related the gravitational potential to the coefficient of dT^2 in the metric, Møller was able to complete the description in the accelerated frame.

It is clear that such treatments (like Tolman's and a further variant due to Fock [103]) can give no value for the time recorded by a clock, between two events on its world-line, other than is found simply by evaluating the proper time $\int ds$ along the path. The only object in introducing an accelerated system into the problem could be to demonstrate the actual observations which the travelling twin might make of the earth twin's clock. But as we have seen, such systems bear little relation to any observations that an accelerated observer M could actually make. An army of helpers distributed throughout space, *and knowing M's intended motion in advance,* would be needed to make local observations for him.

The unreality of these accelerated systems can be shown another way. We know that the frequency of a photon, and hence its energy and mass, vary with the gravitational potential. The law of conservation of energy can be used to show that the mass of any test body must likewise vary with the gravitational potential. When M switches on his rockets, there is a sudden change in the potential in his system, and therefore a sudden change in the inertial mass of a distant test body. Momentum is not changed discontinuously, according to the general theory, and so *there is a jump in the velocity of the test body.* Any reference system in which free test bodies undergo discontinuous changes in velocity is certain to be related in an awkward way to observations.

This velocity discontinuity was noticed in respect of Møller's treatment by Leffert and Donahue [154]. A detailed physical explanation was subsequently given by Møller [182].

We can see that the general theory adds little to the interpretation of the clock paradox in the absence of real gravitational fields, although it is on this case that most disputes have centred. Fortunately, the amount of experimental confirmation of time-dilatation is increasing rapidly. As the phenomenon at last becomes a familiar feature in scientific observations, the 'paradox' of asymmetrical ageing will cease to intrigue or perplex. Whether it will eventually become an important factor in space-travel remains to be seen; but there is surely a chance that it will. . . .

APPENDIX

The relation between the acceleration of a point P moving in the x direction in an inertial system $S(x, t)$, and its acceleration in a second, parallel, system $S'(x', t')$, which is moving with speed V along Ox is obtained as follows.

Let v be the velocity of the point in S, and v' the velocity in S'. By (2.12) and (2.10), we have

$$v = \frac{v' + V}{1 + Vv'/c^2},$$

$$t = \frac{1}{\beta}\left[t' + \frac{Vx'}{c^2}\right],$$

where $\beta = \sqrt{(1 - V^2/c^2)}$. Taking differentials,

$$dv = \frac{dv'}{1 + Vv'/c^2} - \frac{v' + V}{(1 + Vv'/c^2)^2}\left[\frac{V}{c^2}\right]dv$$

$$= \frac{1 - V^2/c^2}{(1 + Vv'/c^2)^2}dv'$$

$$= \frac{\beta^2\, dv'}{(1 + Vv'/c^2)^2},$$

$$dt = \frac{1}{\beta}\left[dt' + \frac{V\, dx'}{c^2}\right]$$

$$= \frac{1}{\beta}dt'\left[1 + \frac{Vv'}{c^2}\right].$$

Dividing the last two equations, we obtain for the acceleration in S,

$$\frac{dv}{dt} = \frac{\beta^3}{(1 + Vv'/c^2)^3}\frac{dv'}{dt'}, \qquad (A.1)$$

where dv'/dt' is the acceleration in S'.

If S' is the instantaneously co-moving system of the point P, then $v' = 0$, $V = v$; and if $dv'/dt' = a$, then (A.1) becomes

$$\frac{dv}{dt} = \beta^3 a, \qquad (A.2)$$

which is equation (3.3).

REFERENCES

(a) The Clock Paradox. Selected Bibliography

[1] ARMSTRONG, H. L. 'Controversy Concerning Time Dilatation', *Amer. J. Phys.*, **28**, 504 (1960). (Pendulum clock in lift.)

[2] ARZELIÈS, H. *Relativistic Kinematics*, Pergamon, Oxford (1966). (Excellent discussion of many issues. Over 100 references to the clock paradox literature.)

[3] ARZELIÈS, H. *Relativité Généralisée, Gravitation*, Fascicules 1 and 2, Gauthier-Villars, Paris (1961 and 1963). (Discusses the clock paradox in general relativity, and gives many references.)

[4] BADESSA, R. S., KENT, R. L. and NOWELL, J. C. 'Short-time Measurement of Time Dilation in an Earth Satellite', *Phys. Rev. Lett.*, **3**, 79 (1959).

[5] BARRETT, W. *Nature*, **216**, 524 (1967).

[6] BENEDIKT, E. T. 'The Clock Paradox in Vertical Free Fall', 7th Annual Meeting of the American Astronautical Society, Preprint 61–43 (1961). (Considers conditions when the 'travelling clock' records the longer time.)

[7] BENTON, MILDRED C. *The Clock Problem (Clock Paradox) in Relativity*, Washington, U.S. Naval Research Laboratory, Bellevue, D.C., Bibliography No. 15 (1959). (Annotated bibliography of nearly 250 articles.)

[8] BERGMANN, O. 'Einige Bemerkungen zum Uhrenparadox', *Acta Phys. Austriaca*, **11**, 377 (1957).

[9] BLASS, G. A. 'On the Clock Paradox in Relativity Theory', *Amer. Math. Monthly*, **67**, 754 (1960).

[10] BOAS, M. L. 'The Clock Paradox', *Science*, **130**, 1471 (1959). (Brief account of geodesic length in Minkowski space-time.)

[11] BONDI, H. 'The Space Traveller's Youth', *Discovery*, **18**, 505 (1957). (Simple account. Introduces k-calculus.)

[12] BORN, M. *Einstein's Theory of Relativity*, Dover, New York (1962). (Contains discussion of clock paradox. 'Thus the clock paradox is due to a false application of the special theory of relativity, namely to a case in which the general theory should be applied.')

[13] BORN, M. 'Special Theory of Relativity', *Nature*, **197**, 1287 (1963). (Reply to H. Dingle [72]. 'Though former experience has taught me that discussing relativity with Dingle leads to no agreement I have to answer a challenge which is directed against the "scientific integrity" of myself and others.')

[14] BOYER, R. H. 'The Clock Paradox in General Relativity', *Nuovo Cim.*, **33**, 345 (1964). (Criticism of [219]. Applies methods of calculus of variations to analyse space-time geodesics.)

[15] BRADBURY, T. C. 'Relativistic Theory of the Behaviour of Clocks', *Amer. J. Phys.*, **28**, 443 (1960). (Frames of reference in arbitrary motion.)

[16] BREWSTER, W. R. 'Life Span Same in Space as on Earth', *Science News Letter* (December 15, 1956), 371.

[17] BROWN, G. B. 'What is wrong with Relativity?', *Bull. Inst. Phys.*, **18**, 71 (1967). ('The device of calling them only apparent contradictions (paradoxes) has not succeeded in preventing the special theory of relativity from becoming untenable as a physical theory.' Editorial summary of criticism, in voluminous correspondence, prepared with assistance of H. Bondi, FRS. and colleagues, and R. E. Peierls, FRS., *ibid.*, **19**, 22 (1968).)

[18] BUILDER, G. 'The Resolution of the Clock Paradox', *Austral. J. Phys.*, **10**, 246 (1957). (Long detailed discussion. Special relativity sufficient.)

[19] BUILDER, G. 'The Clock-retardation Problem', *Austral. J. Phys.*, **10**, 424 (1957). (Defends an earlier paper [17] against attack by H. Dingle [60].)

[20] BUILDER, G. 'The "Clock Paradox"', *Bull. Inst. Phys.*, **8**, 210 (1957). (A reply to H. Dingle [55].)

[21] BUILDER, G. 'Ether and Relativity', *Austral. J. Phys.*, **11**, 279 (1958). ('Accepts' relativity compatible with ether. 'There is no alternative to the ether hypothesis.')

[22] BUILDER, G. 'The Resolution of the Clock Paradox', *Phil. Sci.*, **26**, 135 (1959). (Discussion of clock 'rates'. Nothing paradoxical in predicted retardation. General relativity adds nothing.)

[23] BURCHAM, W. E. 'Nuclear Resonant Scattering without Recoil (The Mössbauer Effect)', *Sci. Prog.*, **48**, 630 (1960). (An account of early applications of the Mössbauer effect.)

[24] CAMPBELL, J. W. 'The Clock Problem in Relativity', *Phil. Mag.*, **15**, 48 (1933). (Approximate treatment involving gravitational potential.)

[25] CAMPBELL, J. W. 'The Clock Problem in Relativity', *Phil. Mag.*, **16**, 529 (1933). (Long paper. Introduces several time-rates in connection with accelerating reference systems.)

[26] CAMPBELL, J. W. 'The Clock Problem in Relativity', *Phil. Mag.*, **19**, 715 (1935). (*Longer* time-interval along some disturbed geodesic paths in Schwarzchild field.)

[27] CAMPBELL, J. W. 'The Nature of Time', *Nature*, **145**, 426 (1940). (Criticism of Dingle's mass and volume clocks [50].)

[28] CAMPBELL, J. W. 'The Clock Paradox', *Can. Aero. J.*, **4**, 316 (1958).

[29] CHAMBADAL, P. *Réalité et Convention*, Appendix: 'Le Voyageur de Langevin', A. Colin, Paris (1960). (Continues a dispute with Arzeliès; see [2], p. 195.)

[30] CHAMPENEY, D. C., ISAAK, G. R. and KHAN, A. M. 'Measurement of Relativistic Time Dilatation using the Mössbauer Effect',
Nature, **198**, 1186 (1963).

[31] CHAMPENEY, D. C., ISAAK, G. R. and KHAN, A. M. 'A Time
Dilatation Experiment based on the Mössbauer Effect', *Proc.
Phys. Soc.*, **85**, 583 (1965).

[32] CHAMPENEY, D. C. and MOON, P. B. 'Absence of Doppler Shift
for Gamma Ray Source and Detector on Same Circular Orbit',
Proc. Phys. Soc., **77**, 350 (1961).

[33] CHAZY, J. *Théorie de la Relativité et Mécanique Céleste*, Vol. 2,
Gauthier-Villars, Paris (1930). (Separated and reunited clocks
record equal elapsed times.)

[34] COCHRAN, W. 'A Suggested Experiment on the Clock Paradox',
Nature, **179**, 977 (1957). (Suggests an improvement to a proposed
experiment of F. S. Crawford [41], which (Cochran maintains)
has already been done.)

[35] COCHRAN, W. 'A Suggested Experiment on the Clock Paradox',
Proc. Camb. Phil. Soc., **53**, 646 (1957). (Elaboration of [34].)

[36] COCHRAN, W. 'The Clock Paradox', *Vistas in Astronomy*, **3**, 78
(1960). (General discussion, and suggestion of π meson experiment.)

[37] COLEMAN, J. A. *Relativity for the Layman*, p. 71, Penguin Books,
London (1959). (Maintains that when two observers are moving
relative to each other the slowing down of time processes applies
only when the relative velocity is constant. There is no permanent
effect in the case of a rocket taking off from earth and later
landing.)

[38] COSTA DE BEAUREGARD, O. 'Relation entre le Temps Propre
d'un Horloge Terrestre et le Temps Astronomique de Schwarzschild à l'Approximation de 10^{-12}', *J. Phys. et Radium*, **18**, 17
(1957). (Considers effects of motion of earth, and masses of earth
and sun on terrestrial clocks.)

[39] COSTA DE BEAUREGARD, O. 'Fragilité des Thèses Antirelati-
vistes de P. Dive et de F. Prunier', *Revue Scientifique* (April 15,
1948), 424. (Part of dispute with P. Dive. See also [79]. For
further references, see [2].)

[40] CRAMPIN, J., MCCREA, W. H. and MCNALLY, D. 'A Class of
Transformations in Special Relativity', *Proc. Roy. Soc.* A, **252**,
156 (1959). (See also [248].)

[41] CRAWFORD, F. S. 'Experimental Verification of the Clock
Paradox of Relativity', *Nature*, **179**, 35 (1957). (Refers to actual
and possible meson experiments. See [34].)

[42] CRAWFORD, F. S. 'The "Clock Paradox" of Relativity', *Nature*,
179, 1071 (1957). (Supports Einstein, not Dingle [59], on matters
of synchronization.)

[43] CROCCO, G. A. Reported in *The Times,* September 19, 1956.

[44] CROWELL, A. D. 'Observations of a Time Interval by a Single Observer', *Amer. J. Phys.,* **29**, 370 (1961).

[45] CULLWICK, E. G. *Electromagnetism and Relativity,* Longmans, London (1957). ('The Lorentz transformation is expressly designed for the case in which the systems start together and move apart, and difficulties occur when the systems are approaching one another.' Concludes that 'both the paradox and its alleged solution are fallacious'.)

[46] CULLWICK, E. G. 'The Clock Paradox', *J. Inst. Elect. Eng.,* **9**, 164 (1963).

[47] DARWIN, C. G. 'The Clock Paradox in Relativity', *Nature,* **180**, 976 (1957). (Standard result via Doppler effect.)

[48] DAVIDSON, W. 'Use of an Artificial Satellite to test the Clock Paradox and General Relativity', *Nature,* **188**, 1013 (1960). (An alternative derivation to one by S. F. Singer [212], of the retardation formula for a satellite clock in circular orbit.)

[49] DINGLE, H. 'Modern Aristotelianism', *Nature,* **139**, 784 (1937).

[50] DINGLE, H. 'The Relativity of Time', *Nature,* **144**, 888 (1939). (Introduces mass and volume clocks on hourglass principle.)

[51] DINGLE, H. 'The Nature of Time', *Nature,* **145**, 427 (1940). (Defends mass and volume clocks [50], against attacks by J. W. Campbell [27].

[52] DINGLE, H. 'The Rate of a Moving Clock', *Nature,* **146**, 391 (1940). (Answers points from 'mass of correspondence' received regarding [50].)

[53] DINGLE, H. 'The Time Concept in Restricted Relativity', *Amer J. Phys.,* **10**, 203 (1942). (Reply to P. S. Epstein's criticism [89] of his book *The Special Theory of Relativity* and hourglass clocks [50].)

[54] DINGLE, H. 'The Time Concept in Restricted Relativity', *Amer. J. Phys.,* **11**, 228 (1943). (Further rejoinder to Epstein [89; 90].)

[55] DINGLE, H. 'What does Relativity mean?', *Bull. Inst. Phys.,* **7**, 314 (1956). (Attacks many others. 'A situation in which the material destiny of the world is in the hands of men manipulating a tool whose nature they quite misconceive is one of extreme peril.')

[56] DINGLE, H. 'Relativity and Space Travel', *Nature,* **177**, 782 (1956). (Answers opponents on matters of simultaneity and clock synchronization.)

[57] DINGLE, H. *Nature,* **177**, 785 (1956). ('McCrea [161] wanders widely from the point, and I do not propose to follow him.')

[58] DINGLE, H. 'Relativity and Space Travel', *Nature,* **178**, 680 (1956). ('Concluding statement' in polemic with McCrea.)

[59] DINGLE, H. 'A Problem in Relativity Theory', *Proc. Phys. Soc. A*, **69**, 925 (1956).

[60] DINGLE, H. 'The Resolution of the Clock Paradox', *Austral, J. Phys.*, **10**, 418 (1957). (Reply to G. Builder [18]. Dynamical asymmetry need not allow one to distinguish 'the *motion* of M from the *motion* of R'.)

[61] DINGLE, H. *Bull. Inst. Phys.*, **8**, 212 (1957). (Further reply to Builder [19].)

[62] DINGLE, H. 'Space Travel and Ageing', *Discovery*, **18**, 174 (April, 1957). (Letter to editor.)

[63] DINGLE, H. *Nature*, **179**, 865 (1957). (Reply to F. S. Crawford [41]. Discusses confusion between coordinate and observed times.)

[64] DINGLE, H. 'The "Clock Paradox" of Relativity', *Nature*, **179**, 1242 (1957). (Further reply to Crawford [42]. No observable phenomenon will show in an absolute sense which of two clocks has moved.)

[65] DINGLE, H. *Nature*, **180**, 500 (1957). (Reply to J. H. Fremlin [106]. The issue of time-at-a-distance.)

[66] DINGLE, H. 'The Clock Paradox in Relativity', *Nature*, **180**, 1275 (1957). (Change in Doppler shift is delayed for accelerated observer. Criticizes Darwin [47].)

[67] DINGLE, H. 'The Interpretation of the Special Relativity Theory', *Bull. Inst. Phys.*, **9**, 314 (1958).

[68] DINGLE, H. 'The Clock Paradox of Relativity', *Science*, **127**, 158 (1958). (Letter to editor. Answers numerous opponents.)

[69] DINGLE, H. 'The Doppler Effect and the Foundations of Physics I'. *Brit. J. Phil. Sci.*, **11**, 11 (1960).

[70] DINGLE, H. 'The Doppler Effect and the Foundations of Physics II', *Brit. J. Phil. Sci.*, **11**, 113 (1960). (Delayed Doppler effect.)

[71] DINGLE, H. 'Relativity and Electromagnetism: An Epistemological Appraisal', *Phil. Sci.*, **27**, 233 (1960). (Different rates of clocks.)

[72] DINGLE, H. 'Special Theory of Relativity', *Nature*, **195**, 985 (1962). (Expresses concern over lack of comment on inconsistency in special relativity, pointed out in [71] and [78]. Serious consequences possible.)

[73] DINGLE, H. 'Special Theory of Relativity', *Nature*, **197**, 1248 (1963). (Answers many replies to [72].)

[74] DINGLE, H. *Nature*, **197**, 1287 (1963). (Reply to Max Born [13], concerning [72].)

[75] DINGLE, H. 'The Case against Special Relativity', *Nature*, **216**, 119 (1967). (Despite attention drawn to error in special relativity five years earlier, 'the theory has continued to be accepted and

used as though it were unquestioned'. Points out 'erroneous conclusion' of Einstein, and gives warning of danger in relying on the theory.)

[76] DINGLE, H. 'The Case against the Theory of Special Relativity', *Nature*, **217**, 19 (1968). (Comments on McCrea's reply [167] to [75].)

[77] DINGLE, H. 'Time in Relativity Theory: Measurement or Coordinate?', in *The Voices of Time* (edited by J. T. Fraser), p. 455, Allen Lane The Penguin Press, London (1968).

[78] DINGLE, H. and SAMUEL, VISCOUNT. *A Threefold Cord*, Allen and Unwin, London (1961).

[79] DIVE, P. 'A propos d'un Article de C. de Beauregard', *Revue Scientifique* (December, 1948), 727. (See [39].)

[80] DURBIN, R., LOAR, H. H. and HAVENS, W. W. 'The Lifetime of the π^+ and π^- Mesons', *Phys. Rev.*, **88**, 179 (1952).

[81] EDDINGTON, SIR ARTHUR S. *Space Time and Gravitation*, University Press, Cambridge (1920).

[82] EDDINGTON, SIR ARTHUR S. *The Nature of the Physical World*, University Press, Cambridge (1928).

[83] EINSTEIN, A. 'Zur Elektrodynamik bewegter Körper', *Ann. d. Phys.*, **17**, 891 (1905). Also as English translation, 'On the Electrodynamics of Moving Bodies', in A. Einstein and others, *The Principle of Relativity*, Methuen, London (1923). (Dover reprint.) (Original formulation of special relativity.)

[84] EINSTEIN, A. 'Über den Einfluss der Schwerkraft auf die Aus breitung des Lichtes', *Ann. d. Phys.*, **35**, 898 (1911). Also as English translation, 'On the Influence of Gravitation on the Propagation of Light', in A. Einstein and others [83]. (Introduction of principle of equivalence.)

[85] EINSTEIN, A. 'Die Grundlage der allgemeinen Relativitäts-theorie', *Ann. d. Phys.*, **49**, 767 (1916). Also as English translation, 'The Foundation of the General Theory of Relativity', in A. Einstein and others [83]. (Formulation of general relativity.)

[86] EINSTEIN, A. 'Dialog über Einwände gegen die Relativitäts-theorie', *Die Naturwiss.*, **6**, 697 (1918). (Resolution of the paradox using general relativity.)

[87] EINSTEIN, A. *Relativity; the Special and the General Theory*, Methuen, London (1920).

[88] EISENLOHR, H. 'Another Note on the Twin Paradox', *Amer. J. Phys.*, **36**, 635 (1968). (Comments on [156].)

[89] EPSTEIN, P. S. 'The Time Concept in Restricted Relativity', *Amer. J. Phys.*, **10**, 1 (1942). (Criticism of H. Dingle, *The Special Theory of Relativity* and [50]. See also [53].)

[90] EPSTEIN, P. S. 'The Time Concept in Restricted Relativity— A Rejoinder', *Amer. J. Phys.*, **10**, 205 (1942). (Answer to [53].

Emphasizes reality of contractions and dilatations. 'Gratuitous buffoonery' on part of Dingle.)

[91]   ESSEN, L. 'The Clock Paradox of Relativity', *Nature,* **180,** 1061 (1957). (Paradox is due to error in Einstein's (1905) paper [83]. Clock has two parts: time standard supplying impulses and dial which counts them. Favours case for symmetrical ageing.)

[92]   ESSEN, L. *Proc. Roy. Soc.* A, **270,** 312 (1962). (Maintains that Einstein, in appendix to [87], uses the same symbol for two different quantities.)

[93]   ESSEN, L. In *Air, Space and Instruments,* edited by S. Lees, McGraw-Hill, New York (1963).

[94]   ESSEN, L. *J. Inst. Elect. Eng.,* **9,** 389 (1963). (Reply to Wilkinson [239] and other correspondents.)

[95]   ESSEN, L. 'Basic Concepts of Measurement and the Michelson–Morley Experiment', *Nature,* **199,** 684 (1963). ('It has been shown [elsewhere] that the well-known clock paradox is predicted by the [special] theory only by changing the meaning of the symbols during the course of the argument.)

[96]   ESSEN, L. 'Bearing of Recent Experiments on the Special and General Theories of Relativity', *Nature,* **202,** 787 (1964). (Criticizes interpretations of Champeney and Moon [32] regarding their rotor experiment using the Mössbauer effect.)

[97]   ESSEN, L. 'A Time Dilatation Experiment based on the Mössbauer Effect', *Proc. Phys. Soc.,* **86,** 671 (1965). (Further attention is drawn to an 'error' in [32]. (See [96].) Also reinterprets an experiment of 'Moon *et al*', but apparently is referring to Champeney, Isaak and Khan [31]. Maintains that authors 'make one assumption which contradicts the special theory and one which contradicts the general theory and their experimental results'.)

[98]   ESSEN, L. 'The Error in the Special Theory of Relativity', *Nature,* **217,** 19 (1968). (According to Essen, McCrea explained the apparent contradiction in Dingle's results by pointing out that Einstein used two symbols for four quantities. Essen had discussed this earlier. Einstein's result follows 'from an assumption made implicitly that the clock which does the round trip is actually going slower than the one regarded as stationary and does not simply appear to be going slower as viewed by the stationary observer.')

[99]   EULER, H. and HEISENBERG, W. 'Theoretische Gesichspunkte zur Deutung der kosmischen Strahlung', *Ergebn. Exact. Naturwiss.,* **17,** 1 (1938). (An early study of the lifetime of μ mesons in flight.)

[100]   FAHY, E. F. 'The Clock Paradox in Relativity', *Austral. J. Phys.,* **11,** 586 (1958). (Comments on Dingle's suggestion of a 'delayed' Doppler effect [66].)

[101] FARLEY, F. J. M., BAILEY, J. and PICASSO, E. 'Experimental Verifications of the Special Theory of Relativity', *Nature*, **217**, 17 (1968).

[102] FISHER, SIR RONALD A. 'Space-travel and Ageing', *Discovery*, **18**, 56 (February, 1957). (Poses some points to McCrea.)

[103] FOCK, V. A. *The Theory of Space Time and Gravitation*, Pergamon, Oxford (1964). (Uses general relativity, without assuming the principle of equivalence.)

[104] FOKKER, A. D. 'Accelerated Spherical Light Wave Clocks in Chronogeometry', *Physica*, **22**, 1279 (1956).

[105] FOKKER, A. D. 'The Clock Paradox in So-called Relativity Theory', *Physica*, **24**, 1119 (1958).

[106] FREMLIN, J. H. 'Relativity and Space Travel', *Nature*, **180**, 499 (1957).

[107] FRYE, R. M. and BRIGHAM, V. M. 'Paradox of the Twins', *Amer. J. Phys.*, **25**, 553 (1957). (Comments in reply to W. R. Brewster of Harvard Medical School.)

[108] FULLERTON, J. H. *Nature*, **216**, 524 (1967). (Letter to editor, in reply to [75].)

[109] GAMBA, A. 'Time Dilatation and Information Theory', *Amer. J. Phys.*, **33**, 61 (1965). (Conventional Doppler effect type of argument.)

[110] GOLAY, M. J. E. 'Note on Relativistic Clock Experiment', *Amer. J. Phys.*, **25**, 495 (1957). (Concerned with clocks at different altitudes.)

[111] GOODHART, C. B. *Discovery*, **18**, 519 (December 1957). (Biological ageing in the clock paradox. But mainly concerned with the effect of temperature on ageing.)

[112] GRÜNBAUM, A. 'The Clock Paradox in the Special Theory of Relativity', *Phil. Sci.*, **21**, 249 (1954).

[113] GUNN, J. A. *The Problem of Time. An Historical and Critical Study*, Allen and Unwin, London (1929).

[114] HALSBURY, LORD. 'Space Travel and Ageing', *Discovery*, **18**, 174 (April, 1957). (3-clock problem.)

[115] HAY, H. J., SCHIFFER, J. P., CRANSHAW, T. E. and EGELSTAFF, P. A. 'Measurement of the Red Shift in an Accelerated System using the Mössbauer Effect in Fe', *Phys. Rev. Lett.*, **4**, 165 (1960).

[116] HILL, E. L. 'On the Kinematics of Uniformly Accelerated Motions and Classical Electromagnetic Theory', *Phys. Rev.*, **72**, 143 (1947).

[117] HILL, E. L. 'The Relativistic Clock Problem', *Phys. Rev.*, **72**, 236 (1947). (Uniformly accelerated motions via conformal transformations. Some non-standard solutions.)

[118] HLAVATY, V. 'Proper Time, Apparent Time and Formal Time in the Twin Paradox', *J. Math. Mech.*, **9**, 733 (1960). (Formal technical discussion from mathematical, rather than physical, viewpoint.)

[119] HOFFMANN, B. 'General Relativistic Red Shift and the Artificial Satellite', *Phys. Rev.*, **106**, 358 (1957). (Effects of earth's diurnal rotation and asymmetry.)

[120] HOFFMANN, B. 'Noon-Midnight Red Shift', *Phys. Rev.*, **121**, 337 (1961).

[121] HOFFMANN, B. and SPROULL, W. T. 'Clock Rates at Perihelion and Aphelion', *Amer. J. Phys.*, **29**, 640 (1961).

[122] HURST, C. A. 'Acceleration and the Clock Paradox', *J. Austral. Math. Soc.*, **2**, 120 (1961). (Investigates case of uniform acceleration, and considers limiting case as acceleration becomes infinite.)

[123] INFELD, L. 'Clocks, Rigid Rods and Relativity Theory', *Amer. J. Phys.*, **11**, 219 (1943). (Written at invitation of editor 'to present dispassionately the differences between Epstein's [89; 90] and Dingle's [53; 54] views'. Emphasizes 'periodic' and 'aperiodic' aspects of clocks.)

[124] ISAAK, G. R. 'The Clock Paradox and the General Theory of Relativity', *Austral, J. Phys.*, **10**, 207 (1957).

[125] IVES, H. E. 'Graphical Exposition of the Michelson–Morley Experiment', *J. Opt. Soc. Amer.*, **27**, 177 (1937).

[126] IVES, H. E. 'Light Signals on Moving Bodies as measured by Transported Rods and Clocks', *J. Opt. Soc. Amer.*, **27**, 263 (1937). (Properties of rods and clocks in motion relative to ether.)

[127] IVES, H. E. 'The Aberration of Clocks and the Clock Paradox', *J. Opt. Soc. Amer.*, **27**, 305 (1937). (Relates properties of transported clocks to stellar aberration.)

[128] IVES, H. E. 'Apparent Lengths and Times in Systems experiencing the Fitzgerald-Larmor-Lorentz Contraction', *J. Opt. Soc. Amer.*, **27**, 310 (1937).

[129] IVES, H. E. 'Derivation of the Lorentz Transformation Equations', *Phil. Mag.*, **36**, 392 (1945).

[130] IVES, H. E. 'Historical Note on the Rate of a Moving Clock', *J. Opt. Soc. Amer.*, **37**, 810 (1947). (Possible variants of Lorentz transformation, and consistency with optical experiments.)

[131] IVES, H. E. 'The Measurement of the Velocity of Light by Signals sent in One Direction', *J. Opt. Soc. Amer.*, **38**, 879 (1948).

[132] IVES, H. E. 'Extrapolation from the Michelson–Morley Experiment), *J. Opt. Soc. Amer.*, **40**, 185 (1950). (One-way measurements of the velocity of light. Time-at-a-distance assumed largely to be indeterminate. Obtains Lorentz transformation with preferred 'isotropic' coordinate system.)

[133] IVES, H. E. 'The Clock Paradox in Relativity Theory', *Nature*, **168**, 246 (1951).

[134] IVES, H. E. and STILWELL, G. R. 'An Experimental Study of the Rate of a Moving Atomic Clock', *J. Opt. Soc. Amer.*, **28**, 215 (1938). (Second order Doppler effect in canal ray sources.)

[135] IVES, H. E. and STILWELL, G. R. 'An Experimental Study of the Rate of a Moving Atomic Clock II', *J. Opt. Soc. Amer.*, **31**, 369 (1941). (Continuation of experiments in [134], with additional refinements. Deals with points raised by Jones [138].)

[136] JAYNES, E. T. 'Relativistic Clock-experiments', *Amer. J. Phys.*, **26**, 197 (1958).

[137] JEFFREYS, H. 'The Clock Paradox in Special Relativity', *Austral. J. Phys.*, **11**, 583 (1958). (Compares analyses of Builder [18] and Dingle [60], and concludes that special relativity does not give a unique answer.)

[138] JONES, R. C. 'On the Relativistic Doppler Effect', *J. Opt. Soc. Amer.*, **29**, 337 (1939). (Discusses theory of Ives–Stilwell experiment. See [135].)

[139] JOSEPHSON, B. D. 'Temperature-dependent Shift of X-Rays emitted by a Solid', *Phys. Rev. Lett.*, **4**, 341 (1960).

[140] KENNEDY, R. J. and THORNDIKE, E. M. 'Experimental Establishment of the Relativity of Time', *Phys. Rev.*, **42**, 400 (1932).

[141] KERMACK, W. O., MCCREA, W. H. and WHITTAKER, E. T. 'On Properties of Null Geodesics and the Application to the Theory of Radiation', *Proc. Roy. Soc. Edin.*, **53**, 31 (1933).

[142] KOPFF, A. *The Mathematical Theory of Relativity*, Methuen, London (1923). (Uses principle of equivalence and apparent gravitational field for accelerated observer.)

[143] KOWALSKI, K. L. 'Relativistic Reaction Systems and the Asymmetry of Time-scales', *Amer. J. Phys.*, **28**, 487 (1960).

[144] KRONSBEIN, J. and FARBER, E. A. 'Time Retardation in Static and Stationary Spherical and Elliptic Spaces', *Phys. Rev.*, **115**, 763 (1959).

[145] KÜNDIG, W. 'Measurement of the Transverse Doppler Effect using the Mössbauer Effect', *Bull. Amer. Phys. Soc.*, **7**, 350 (1962). (Description of work in progress.)

[146] KÜNDIG, W. 'Measurement of the Transverse Doppler Effect in an Accelerated System', *Phys. Rev.*, **129**, 2371 (1963). (Description of rotor experiment based on Mössbauer effect.)

[147] KURONOMA, E. 'A New Solution of the Clock Paradox', *Prog. Theor. Phys. Japan*, **25**, 508 (1961). (Doppler effect and simultaneity argument.)

[148] KUTLIROFF, D. 'Time Dilatation Derivation', *Amer. J. Phys.*, **31**, 137 (1963).

[149] LANDSBERG, P. T. *Nature,* **220,** 1182 (1968). (Reply to H. Dingle [76].)

[150] LANGE, L. 'The Clock Paradox in the Theory of Relativity', *Amer. Math. Monthly,* **34,** 22 (1927). (Queries validity of clock hypothesis.)

[151] LANGEVIN, P. 'l'Evolution de l'Espace et du Temps', *Scientia,* **10,** 31 (1911). (Classic early paper, giving detailed exposition.)

[152] LARMOR, J. *Aether and Matter,* C.U.P., London (1900).

[153] LASS, H. 'Accelerated Frames of Reference and the Clock Paradox', *Amer. J. Phys.,* **31,** 274 (1963).

[154] LEFFERT, C. B. and DONAHUE, T. M. 'Clock Paradox and the Physics of Discontinuous Gravitational Fields', *Amer. J. Phys.,* **26,** 514 (1958). (Shows that sudden changes in gravitational fields, such as occur in Møller's general relativistic treatment [179], cause discontinuities in the velocity of the inertial clock in the accelerated rest system of the other clock.)

[155] LEROUX, J. 'Sur une Forme Nouvelle des Formules de Lorentz', *C. R. Acad. Sci. Paris,* **197,** 394 (1933).

[156] LEVI, L. 'The Twin Paradox Revisited', *Amer. J. Phys.,* **35,** 968 (1967).

[157] LOWRY, E. S. 'The Clock Paradox', *Amer. J. Phys.,* **31,** 59 (1963).

[158] MCCREA, W. H. 'The Clock Paradox in Relativity Theory', *Nature,* **167,** 680 (1951). (Treatment in terms of distance variation due to Fitzgerald–Lorentz contraction.)

[159] MCCREA, W. H. 'The Fitzgerald–Lorentz Contraction – Some Paradoxes and their Solution', *Sci. Proc. Roy. Dublin Soc.,* **26,** 27 (1952).

[160] MCCREA, W. H. 'A Time-keeping Problem connected with Gravitational Red Shift', *Helv. Phys. Acta Suppl.,* **4,** 121 (1956). (Circular orbits in central gravitational field.)

[161] MCCREA, W. H. 'Relativity and Space Travel', *Nature,* **177,** 784 (1956). (Reply to Dingle [56]. Emphasizes asymmetry of space-time routes. 'Dingle's "paraphrase" of Einstein's paper is a travesty.')

[162] MCCREA, W. H. 'Relativity and Space Travel', *Nature,* **178,** 681 (1956). ('Concluding statement' in polemic with Dingle; see [58]. Answers several points raised by correspondents regarding inertial frames, accelerations, etc.)

[163] MCCREA, W. H. 'A Problem in Relativity Theory: Reply to H. Dingle', *Proc. Phys. Soc.* A, **69,** 935 (1956). (Reply to [59].)

[164] MCCREA, W. H. *Discovery,* **18,** 57 (February, 1957). (Answer to Fisher [102].)

[165] MCCREA, W. H. *Discovery,* **18,** 175 (April, 1957), (Letter to editor, answering several correspondents.)

[166] MCCREA, W. H. 'Relativistic Ageing', *Nature*, **179**, 909 (1957). (The 'principle of impotence'.)

[167] MCCREA, W. H. 'Why the Special Theory is Correct', *Nature*, **216**, 122 (1967). (Answer to [75].)

[168] MACDUFFEE, C. C. 'Arc Lengths in Special Relativity', *Proc. Camb. Phil. Soc.*, **56**, 176 (1960).

[169] MCMILLAN, E. M. 'The Clock Paradox and Space Travel', *Science*, **126**, 381 (1957). (Demonstrates asymmetry, introducing accelerated reference frame. Considers practicality of long-distance space-travel.)

[170] MCMILLAN, E. M. *Science*, **127**, 160 (1958). (Reply to criticism by Dingle [68] of his paper [169].)

[171] MCVITTIE, G. C. 'Remarks on Planetary Theory in General Relativity', *Astronom. J.*, **63**, 448 (1958). ('There is a sense in which both Dingle and McCrea are right according to the different permissible definitions of the physiological times by which the twins live.')

[172] MARITAIN, J. 'Einstein et la Notion du Temps', *Revue Univ.* (July, 1920), 358.

[173] MARTINELLI, E. and PANOFSKY, W. K. H. 'The Lifetime of the Positive π Meson', *Phys. Rev.*, **77**, 465 (1950).

[174] METZ, A. *La Relativité*, Chiron, Paris (1923).

[175] METZ, A. 'Une Définition Relativiste de la Simultanéité', *C. R. Acad. Sci. Paris*, **180**, 1827 (1925).

[176] MIKHAIL, F. I. 'The Relativistic Clock Problem', *Proc. Camb. Phil. Soc.*, **48**, 608 (1952). (Example of free clocks separating and reuniting in Schwarzschild space-time, and registering different elapsed times.)

[177] MILNE, E. A. and WHITROW, G. J. 'On the So-called "Clock Paradox" of Special Relativity', *Phil. Mag.*, **40**, 1244 (1949). (There is no paradox in the asymmetrical result on the basis of kinematical relativity.)

[178] MØLLER, C. 'On Homogeneous Gravitational Fields in the General Theory of Relativity', *Det. kgl. danske vid. selsk. mat-fys. meddr.*, **20**, No. 19 (1943).

[179] MØLLER, C. *The Theory of Relativity*, Clarendon, Oxford (1952). (Solution via general relativity; similar to [178].)

[180] MØLLER, C. 'Old Problems in the General Theory of Relativity viewed from a New Angle', *Det. kgl. danske vid. selsk. mat-fys. meddr.*, **30**, No. 10 (1955). (Dynamics of 'ideal' standard clock mechanisms.)

[181] MØLLER, C. 'On the Possibility of Terrestrial Tests of the General Theory of Relativity', *Nuovo Cim.*, **6**, Suppl. 1, 381 (1957). (Satellite clocks.)

[182] MØLLER, C. 'Motion of Free Particles in Discontinuous Gravitational Fields', *Amer. J. Phys.*, **27**, 491 (1959). (Physical explanation of findings in [154].)

[183] 'News and Views', *Nature*, **216**, 113 (1967). (Editorial in connection with [75] and [167].)

[184] *The Observer* (1956). ('Can Space Travel Postpone Death?', by John Davy, 29 April. Numerous letters to editor, 6, 13, and 20 May.)

[185] PAGE, L. 'A New Relativity', *Phys. Rev.*, **49**, 254 (1936). ('Equivalent' relatively accelerated frames, following kinematical ideas of Milne.)

[186] PALACIOS, J. 'The Relativistic Behaviour of Clocks', *Rev. Acad. Ci. Madrid*, **56**, 287 (1962). (There are many papers by this author, who believes that the clock paradox indicates the need for a revision of relativity theory. Further references will be found in [2].)

[187] 'Paradox Persists, The', *J. Inst. Elect. Eng.*, **9**, 459 (1963). (Editorial discussion of 'the unprecedented volume of correspondence on the clock paradox' received in 1963. Some further letters appear in various issues in 1964; the journal was retitled *Electronics and Power* in that year.)

[188] PIERCE, J. R. 'Relativity and Space Travel', *Proc. Inst. Radio Eng.*, **47**, 1053 (1959). (Doppler shift argument showing asymmetry. See also p. 1778, for correspondence on this article.)—

[189] PILGERAM, L. O. 'Time Dilatation', *Science*, **138**, 1180 (1962). (Von Hoerner [231] 'makes unwarranted assumptions when he attempts to apply Einstein's relativity theory to biological time'.)

[190] POUND, R. V. and REBKA, Jr., G. A. 'Variation with Temperature of the Energy of Recoil-free Gamma Rays from Solids', *Phys. Rev. Lett.*, **4**, 274 (1960).

[191] POUND, R. V. and REBKA, Jr., G. A. 'Apparent Weight of Photons', *Phys. Rev. Lett.*, **4**, 337 (1960). (Gravitational frequency shift test using Mössbauer effect.)

[192] PROELL, W. 'Relativity and Space Travel', *J. Space Flight*, **1**, 8 (1949).

[193] PROKHOVNIK, S. J. *The Logic of Special Relativity*, C.U.P., London (1967).

[194] RAPIER, P. M. 'A Proposed Test of the Asymmetrical Ageing Absurdity using Clock Satellites', *Rev. Acad. Madrid*, **57**, 77 (1963).

[195] RASETTI, F. 'Mean Life of Slow Mesotrons', *Phys. Rev.*, **59**, 613 (1941).

[196] ROBINSON, J. D. and FEENBERG, E. 'Time Dilatation and the Doppler Effect', *Amer. J. Phys.*, **25**, 490 (1957).

[197] ROMAIN, J. E. 'Time Measurement in Accelerated Frames of Reference', *Rev. Mod. Phys.*, **35**, 376 (1963). (Comprehensive discussion of accelerated frames in flat space-time. Assesses alternative approaches via clock hypothesis and by assumption of invariable vacuum speed of light.)

[198] ROMER, R. H. 'Twin Paradox in Special Relativity', *Amer. J. Phys.*, **27**, 131 (1959).

[199] ROSSER, W. F. V. *An Introduction to the Theory of Relativity*, Butterworths, London (1964). (Critique of clock paradox, with experimental bias.)

[200] ROSSI, B. and HALL, D. B. 'Variation of the Rate of Mesotrons with Momentum', *Phys. Rev.*, **59**, 223 (1941).

[201] ROSSI, B., HILBERRY, N. and HOAG, J. B. 'The Variation of the Hard Component of Cosmic Rays with Height and the Disintegration of Mesotrons', *Phys. Rev.*, **57**, 461 (1940).

[202] ROWLAND, E. N. 'A Note on Space Travel in a Gravitational Field', *J. Brit. Interpl. Soc.*, **16**, 216 (1957).

[203] SÄNGER, E. *The Attainability of the Stars*, Rand Corporation, Santa Monica, Calif. (1956). (Paper read at Seventh International Astronautical Congress. Time-dilatation and rocketry.)

[204] SÄNGER, E. Reported in *The Sunday Times*, (September 23, 1956). (Account of paper [203]. 'Atom powered space ships now in realms of possibility, which could approach speed of light.')

[205] SÄNGER, E. 'Flight Mechanics of Photon Rockets', *Aero Dig.*, **73**, 68 and 72 (1956).

[206] SCHILD, A. 'The Clock Paradox in Relativity Theory', *Amer. Math. Monthly*, **66**, 1 (1959). (This is one of the most lucid expositions in the literature.)

[207] SCHLEGEL, R. 'New Clock Problems in Special Relativity', *Bull. Am. Phys. Soc.*, **2**, 239 (1957).

[208] SCHLEGEL, R. *Time and the Physical World*, Michigan State Univ. Press, East Lansing, Mich. (1961).

[209] SCOTT, G. D. 'On Solutions of the Clock Paradox', *Amer J. Phys.* **27**, 580 (1959).

[210] SEARS, F. W. *Amer. J. Phys.*, **32**, 570 (1964). (Brief note. Is the correct word 'dilation' or 'dilatation'? Suggests latter preferable.)

[211] SHERWIN, C. W. 'Some Recent Experimental Tests of the Clock Paradox', *Phys. Rev.*, **120**, 17 (1960).

[212] SINGER, S. F. 'Application of an Artificial Satellite to the Measurement of the General Relativistic Red Shift', *Phys. Rev.* **104**, 11 (1956).

[213] SINGER, S. F. 'Space Vehicles as Tools for Research in Relativity', *J. Astronautics*, **4**, No. 3, 49 (1957).

[214] SINGER, S. F. 'Relativity and Space Travel', *Nature,* **179,** 977 (1957). (Concurs with McCrea [161; 162]. Suggests satellite experiment.)

[215] SOKOLOV, A. A. 'The Clock Paradox in the Motion of Charged Particles in a Magnetic Field', *Sov. Phys. Dokl.,* **5,** 287 (1960). (Explicit calculation.)

[216] STEHLING, K. R. 'Space Travel and Relativity or How to Keep from Growing Old', *Jet Propul.,* **26,** 1105 (1956). (Reviews several arguments in the literature.)

[217] SUBOTOWICZ, M. 'Satellites for checking Einstein's Relativity Theory', *Missiles and Rockets,* **2,** 57 (1957). (Retardation of a satellite clock in a year is measurable.)

[218] SWANN, W. F. G. 'Certain Matters in relation to the Restricted Theory of Relativity, with Special Reference to the Clock Paradox and the Paradox of the Identical Twins. I. Fundamentals,' *Amer. J. Phys.,* **28,** 55 (1960). 'II. Discussion of the Problem of the Identical Twins', **28,** 319 (1960).

[219] TANGHERLINI, F. R. 'Postulational Approach to Schwarzschild's Exterior Solution with Application to a Class of Interior Solutions', *Nuovo Cim.,* **25,** 1081 (1961).

[220] TAYLOR, N. W. 'Note on the Harmonic Oscillator in General Relativity', *J. Austral. Math. Soc.,* **2,** 206 (1961). (Oscillations about centre in Schwarzschild's interior solution.)

[221] TERRELL, J. 'Relativistic Observations and the Clock Problem', *Nuovo Cim.,* **16,** 457 (1960).

[222] THIRUVENKATACHAR, V. R. 'Relativity and Space Travel', *Current Science,* **27,** No. 9, 327 (1958).

[223] THOMSON, SIR GEORGE, *The Foreseeable Future,* University Press, Cambridge (1955).

[224] TICHO, H. 'The Mean Life of Mesons at an Altitude of 11,500 ft.', *Phys. Rev.,* **72,** 255 (1947).

[225] TICHO, H. and SCHEIN, M. 'The Mean Life of Negative Mesotrons in Sodium Fluoride', *Phys. Rev.,* **73,** 81 (1948).

[226] 'Time Out', *Scientific American,* **195,** No. 6, 58 (December, 1956). (Editorial on the Dingle-McCrea controversy.)

[227] TOLMAN, R. C. *Relativity, Thermodynamics and Cosmology,* Clarendon, Oxford (1934). (Approximate 'solution' in general relativity, using principle of equivalence.)

[228] TORNEBOHM, H. 'The Clock Paradox and Notion of Clock Retardation in the Special Theory of Relativity', *Theoria (Lund),* **29,** 79 (1963).

[229] VOIGT, W. 'Über das Doppler'sche Princip.', *Ges. Wiss. Göttingen Nachr.,* **10,** 41 (1887). (Quoted by Ives [130] as the first suggestion that a 'natural' clock would alter its rate on motion.)

[230] VON HOERNER, S. 'The Search for Signals from Other Civilizations', *Science*, **134**, 1839 (1961).

[231] VON HOERNER, S. 'The General Limits of Space Travel', *Science*, **137**, 18 (1962).

[232] VON HOERNER, S. 'Time Dilatation', *Science*, **138**, 1180 (1962). (Letter in reply to Pilgeram [189].)

[233] VON KRZYWOBLOCKI, M. Z. 'Time-dilatation Dilemma and Scale Variation', *A.I.A.A.J.*, **2**, 2213 (1964). (Accepts hypothesis of R. Schlegel, that macroscopic thermodynamic processes ('Clausius' processes) are time-invariant and independent of the relativistic transformations, in contrast to 'Lorentz' processes. Requires clocks following different time scales for different purposes.)

[234] VON L'AUE, M. *Die Relativitätstheorie*, Vieweg, Braunschweig (1952).

[235] WESTON, B. *Discovery*, **18**, 174 (April, 1957). (Letter to editor. Answered by McCrea [165].)

[236] WHITEMAN, M. *The Philosophy of Space and Time*, Allen and Unwin, London (1967). ('When we call a theory *ad hoc* the general impression at the back of our mind is that it is "artificial" in some way . . . lacking in substance. . . . To make our judgements as definite as possible we may allot points for *ad hocness* to each theory, out of a maximum of 5; for example (first column):
Contraction and Time Dilatation (today):   5   0
Ptolemaic Epicycles AD 200:   4   1

.   .   .   .   .

.   .   .   .
Arithmetic, in counting objects: 0   5
The second column . . . may be usefully defined as the Conventional Index of Scientific Character.')

[237] WHITROW, G. J. *The Natural Philosophy of Time*, Nelson, London and Edinburgh (1961).

[238] WHITTAKER, E. *From Euclid to Eddington. A Study of Conceptions of the External World*, C.U.P., Cambridge (1949).

[239] WILKINSON, K. J. R. 'An Analysis of the Clock Paradox', *J. Inst. Elect. Eng.*, **9**, 10 (1963). (Orthodox discussion involving simultaneity lines. Rejects arguments of Cullwick and Dingle. Further comments and correspondence: *ibid*, **9**, 165 (1963); **9**, 217 (1963).)

[240] WINTERBERG, F. 'Relativistische Zeitdilation eines Kuenstlichen Satelliten', *Astronautica Acta*, **2**, 25 (1956). (Translated title: 'Relativistic Time Dilatation in an Artificial Satellite'. Slowing of satellite clocks, according to general relativity, is measurable.)

[241] WITTEN, L. 'Experimental Aspects of General Relativity Theory', *J. Astronautics*, **4**, No. 3, 46 (1957).

(b) Other References

[242] ADAMS, W. *Proc. Nat. Acad.*, **11**, 382 (1925).

[243] ALLEN, C. W. *Astrophysical Quantities*, Athlone Press, London (1963).

[244] ALVÄGER, T., FARLEY, F. J. M., KJELLMAN, J. and WALLIN, I. *Phys. Lett.*, **12**, 260 (1964).

[245] BJORKLUND, R., CRANDALL, W. E., MOYER, B. J. and YORK, H. F. *Phys. Rev.*, **77**, 213 (1950).

[246] BOHM, D. *The Special Theory of Relativity*, Benjamin, New York (1965).

[247] BONDI, H. *Cosmology*, University Press, Cambridge (1961).

[248] BORN, M. and BIEM, W. *Proc. Acad. Sci. Amst.* B, **61**, 110 (1958).

[249] BROWN, F. A. 'Living Clocks', *Science*, **130**, 1535 (1959).

[250] BROWN, F. A. 'Biological Clocks', BSCS Pamphlet No. 2, *Amer. Inst. of Biol. Sciences*, Heath, Boston (1962).

[251] BUILDER, G. 'The Constancy of the Velocity of Light', *Austral. J. Phys.*, **11**, 458 (1958).

[252] BÜNNING, E. *The Physiological Clock*, Academic Press, New York (1964).

[253] CLOUDSLEY-THOMSON, J. L. 'Time Sense of Animals', in *The Voices of Time*, edited by J. T. Fraser, Allen Lane The Penguin Press, London (1966).

[254] DE SITTER, W. *Proc. Acad. Sci. Amst.*, **15**, 1297 (1913); **16**, 395 (1913).

[255] DICKE, R. H. 'Experimental Relativity', in *Relativity, Groups and Topology*, edited by de Witt and de Witt, Blackie, London and Glasgow (1964). See also DICKE, R. H. 'The Eötvös Experiment', *Scientific American*, **205**, No. 6, 92 (1961); and *The Theoretical Significance of Experimental Relativity*, Blackie, London and Glasgow (1964).

[256] DYSON, F. J. 'Interstellar Transport', *Physics Today*, **21**, 41 (October, 1968).

[257] EINSTEIN, A. 'Autobiographical Notes', in *Albert Einstein: Philosopher Scientist*, 2nd ed., edited by P. A. Schilpp, Tudor, New York (1951).

[258] EÖTVÖS, BARON ROLAND V. *Math. u. Naturw. Bers. aus Ungarn*, **8**, 65 (1890).

[259] EÖTVÖS, R. V., PEKAR, D. and FEKETE, E. *Ann. d. Phys.*, **68**, 11 (1922).

[260] FLAVELL, J. H. *The Developmental Psychology of Jean Piaget*, Van Nostrand Reinhold, Princeton (1963).

[261] FRAISSE, P. *The Psychology of Time*, Harper and Row, New York (1963).

[262] FRENCH, A. P. *Special Relativity*, Nelson, London (1968).

[263] GRÜNBAUM, A. *Amer. J. Phys.*, **23**, 450 (1955).

[264] HAMNER, K. C. 'Experimental Evidence for the Biological Clock', in *The Voices of Time*, edited by J. T. Fraser, Allen Lane The Penguin Press, London (1966).

[265] HAMNER, K. C. 'Endogenous Rhythms in Controlled Environments', in *Environmental Control of Plant Growth*, Academic Press, New York (1963).

[266] HESSE, M. B. *Forces and Fields*, Nelson, London (1961).

[267] HOFFMANN, B. *Phys. Rev.*, **121**, 337 (1961).

[268] HOLTON, G. In *Relativity Theory: Its Origins and Impact on Modern Thought*, edited by L. Pearce Williams, Wiley, New York (1968).

[269] HOYLE, F. *Frontiers of Astronomy*, Heinemann, London (1955).

[270] JAFFE, B. *Michelson and the Speed of Light*, Heinemann, London (1961).

[271] JAKOBSON, M., SCHULTZ, A. and STEINBERGER, J. *Phys. Rev.*, **81**, 894 (1951).

[272] JEANS, SIR JAMES. *The Mysterious Universe*, University Press, Cambridge (1930).

[273] LATTES, C. M. G., MUIRHEAD, H., OCCHIALINI, G. P. S. and POWELL, C. F. *Nature*, **159**, 694 (1947).

[274] LODGE, SIR OLIVER J. *Nature*, **46**, 165 (1892).

[275] LOVELL, SIR BERNARD. *The Individual and the Universe*, BBC Reith Lectures 1958, published in book form by Oxford University Press, London (1959).

[276] LOVELL, SIR BERNARD. *The Exploration of Outer Space*, Oxford University Press, London (1962).

[277] LORENTZ, H. A. *Proc. Acad. Sci. Amst.*, **6**, 809 (1904). Reprinted in Einstein, A. and others (*The Principle of Relativity*, Methuen, London (1923). (Also Dover reprint.)

[278] MARDER, L. *An Introduction to Relativity*, Longman, London and Harlow (1968).

[279] MENZEL, D. H., BHATNAGER, P. L. and SEN, H. K. *Stellar Interiors*, Chapman and Hall, London (1963).

[280] MICHELSON, A. A. *Amer. J. Sci.* (3), **22**, 20 (1881).

[281] MICHELSON, A. A. and MORLEY, E. W. *Amer. J. Sci.* (3), **34**, 333 (1887).

[282] MILLER, D. C. *Rev. Mod. Phys.*, **5**, 203 (1933). See also the explanation of Miller's result given by Shankland *et al*, *Rev. Mod. Phys.*, **27**, 157 (1955).

[283] MILLER, G. A. and TAYLOR, W. G. *J. Acoust. Soc. Am.*, **20**, 171 (1948).

[284] MILNE, E. A. *Kinematical Relativity*, Clarendon, Oxford (1948).

[285] MINKOWSKI, H. 'Space and Time', Translation of an Address delivered at the 80th Assembly of German Natural Scientists and Physicians, Cologne, 21 September 1908. In Einstein, A. and others, *The Principle of Relativity*, Methuen, London (1923). (Dover reprint.)

[286] MÖSSBAUER, R. L. *Zeit Phys.*, **151**, 124 (1958); *Naturwiss.*, **45**, 538 (1958).

[287] NERESON, N. and ROSSI, B. *Phys. Rev.*, **64**, 199 (1943).

[288] PIAGET, J. *Le Développement de la Notion de Temps chez l'Enfant*, Presses Universitaire de France, Paris (1946).

[289] POINCARÉ, H. *Bull. des Sc. Math.* (2), **28**, 302 (1904). An English translation appears in *The Monist* (January, 1905).

[290] RINDLER, W. *Special Relativity*, Oliver and Boyd, Edinburgh (1960).

[291] ROBB, A. A. *Geometry of Space and Time*, 2nd ed., Macmillan, New York (1936); *A Theory of Space and Time*, C.U.P., London (1914).

[292] ROBERTSON, H. P. *Rev. Mod. Phys.*, **21**, 378 (1949). (See also the axiomatic treatment of Schwartz [296].)

[293] ROSSI, H., HILBERRY, N. and HOAG, J. B. *Phys. Rev.*, **57**, 461 (1940).

[294] SADLER, D. H. *Quart, J. Roy. Ast. Soc.*, **9**, 281 (1968).

[295] ST. JOHN, C. E. *Astrophys. J.*, **67**, 195 (1928); ADAMS, W. S. *Proc. Nat. Acad.*, **11**, 382 (1925).

[296] SCHWARTZ, H. M. *Amer. J. Phys.*, **30**, 697 (1962).

[297] SINGH, JAGJIT. *Great Ideas and Theories of Modern Cosmology*, Constable, London (1961).

[298] SMITH, A. U. *Biological Effects of Freezing and Supercooling*, Arnold, London (1961).

[299] SOUTHERNS, L. *Proc. Roy. Soc.* A, **84**, 325 (1910).

[300] SPENCER JONES, H. *Mon. Not. Roy. Ast. Soc.*, **99**, 541 (1939).

[301] SPENCER JONES, H. *General Astronomy*, 4th ed., Edward Arnold, London (1961).

[302] WHITTAKER, SIR EDMUND T. 'Some Disputed Questions in the Philosophy of the Physical Sciences', *Phil. Mag.* (7), **33**, 353 (1942).

[303] WHITTAKER, SIR EDMUND T. *History of the Theories of Aether and Electricity: The Classical Theories*, Nelson, London (1951).

[304] WHITTAKER, SIR EDMUND T. *History of the Theories of Aether and Electricity: The Modern Theories (1920-1926)*, Nelson, London (1953).

[305] WOODROW, H. In *Handbook of Experimental Psychology*, edited by S. S. Stevens, Wiley, New York (1951).

INDEX